Hydropower

Hydropower

Royce Shaw

Callisto Reference

www.callistoreference.com

Callisto Reference,
118-35 Queens Blvd., Suite 400,
Forest Hills, NY 11375, USA

Visit us on the World Wide Web at:
www.callistoreference.com

ISBN: 978-1-64116-589-1 (Hardback)

Cataloging-in-Publication Data

Hydropower / Royce Shaw.
p. cm.
Includes bibliographical references and index.
ISBN 978-1-64116-589-1
1. Water-power. 2. Renewable energy sources. 3. Power resources.
4. Water resources development. I. Shaw, Royce.

TC147 .H93 2022
333.914--dc23

Table of Contents

Preface

This book aims to help a broader range of students by exploring a wide variety of significant topics related to this discipline. It will help students in achieving a higher level of understanding of the subject and excel in their respective fields. This book would not have been possible without the unwavered support of my senior professors who took out the time to provide me feedback and help me with the process. I would also like to thank my family for their patience and support.

Hydropower refers to the power that is obtained from the energy of falling and fast running water. It is a renewable energy source which can be used for irrigation and the operation of various mechanical devices including gristmills, sawmills, trip hammers, textile mills, domestic lifts and ore mills. It is also used to produce hydroelectricity. Hydroelectric power plants can have a reservoir through which the energy of falling water can be converted into electricity. The kinetic energy of running water can also be converted into electricity through run-of-the-river hydroelectricity generation plants. Another form of electricity generation through hydropower is through tidal stream generators which make use of tidal power from oceans and rivers. This book presents the complex subject of hydropower in the most comprehensible and easy to understand language. It provides comprehensive insights into this field. This book is an essential guide for both academicians and those who wish to pursue this discipline further.

A brief overview of the book contents is provided below:

Chapter - Introduction

The power which is derived from the energy of falling or fast-funning water is referred to as hydropower. Some of the major concepts within this field are low head hydropower and hydraulic head. The chapter closely examines these key concepts of hydropower to provide an extensive understanding of the subject.

Chapter - Hydroelectricity and its types

The electricity produced by generators which are pushed by the movement of water is termed as hydroelectricity. It can be produced at various scales which are generally classified as small hydro, micro hydro and pico hydro. These types of hydroelectricity have been thoroughly discussed in this chapter.

Chapter - Hydroelectricity Generation

Generation of hydroelectricity generally takes place in hydroelectric power plants, which convert the kinetic energy of water into electricity. Some of the major components of hydroelectric power plants are dam, turbine and generator. This chapter has been carefully written to provide an easy understanding of these facets of hydroelectricity generation.

Chapter - Hydroelectricity Equipments

The main equipment used for producing hydroelectricity are water turbines and water wheels. Some of the common types of water turbines are Francis Turbine, Kaplan Turbine, Gorlov Helical Turbine, Pelton Turbine, Turgo Turbine and Cross-flow Turbine. This chapter discusses these hydroelectricity equipment in detail.

Chapter - Hydropower Plants

A few of the common types of hydropower plants are conventional hydroelectric power stations, pumped-storage hydroelectric power stations, run-of-the-river hydroelectric power stations and tidal power stations. The topics elaborated in this chapter will help in gaining a better perspective about these types of hydropower plants.

Chapter - Tidal Power

The form of hydropower that is used for the conversion of energy derived from tides into useful forms of power is referred to as tidal power. Some of the methods of generating tidal power are tidal stream generator, tidal barrage, dynamic tidal power and tidal lagoon. This chapter has been carefully written to provide an easy understanding of these methods of tidal power generation.

Royce Shaw

1

Introduction

The power which is derived from the energy of falling or fast-funning water is referred to as hydropower. Some of the major concepts within this field are low head hydropower and hydraulic head. The chapter closely examines these key concepts of hydropower to provide an extensive understanding of the subject.

HYDROPOWER

Hydropower is energy derived from flowing water. More than 2,000 years ago, the ancient Greeks used waterpower to run wheels for grinding grain; today it is among the most cost-effective means of generating electricity and is often the preferred method where available. In Norway, for example, 99% of electricity comes from hydropower. The world's largest hydropower plant is the 22.5 gigawatt Three Gorges Dam in China. It produces 80 to 100 terawatt-hours per year, enough to supply between 70 million and 80 million households.

Small-scale micro-hydropower projects can make a big difference to communities in remote locations.

The basic principle of hydropower is using water to drive turbines. Hydropower plants consist of two basic configurations: with dams and reservoirs, or without. Hydropower dams with a large reservoir can store water over short or long periods to meet peak demand. The facilities can also be divided into smaller dams for different purposes, such as night or day use, seasonal storage, or pumped-storage reversible plants, for both pumping and electricity generation. Hydropower without dams and reservoirs means producing at a smaller scale, typically from a facility designed to operate in a river without interfering in its flow. For this reason, many consider small-scale hydro a more environmentally-friendly option.

People have a long history of using the force of water flowing in streams and rivers to produce mechanical energy. Hydropower was one of the first sources of energy used for electricity generation and is the largest single renewable energy source for electricity generation in the United States.

In 2018, hydroelectricity accounted for about 7% of total U.S. utility-scale electricity generation and 41% of total utility-scale electricity generation from renewable energy sources. Hydroelectricity's share of total U.S. electricity generation has decreased over time, mainly because electricity generation from other sources has increased.

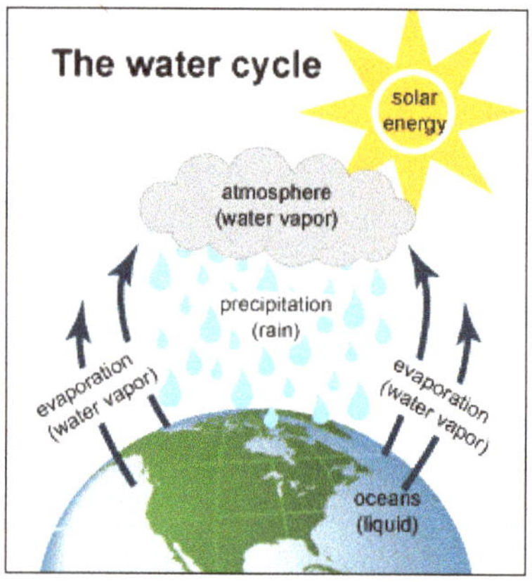

Hydropower Relies on the Water Cycle

Understanding the water cycle is important to understanding hydropower. The water cycle has three steps:

- Solar energy heats water on the surface of rivers, lakes, and oceans, which causes the water to evaporate.
- Water vapor condenses into clouds and falls as precipitation—rain and snow.
- Precipitation collects in streams and rivers, which empty into oceans and lakes, where it evaporates and begins the cycle again.

The amount of precipitation that drains into rivers and streams in a geographic area determines the amount of water available for producing hydropower. Seasonal variations in precipitation and long-term changes in precipitation patterns, such as droughts, have a big impact on hydropower production.

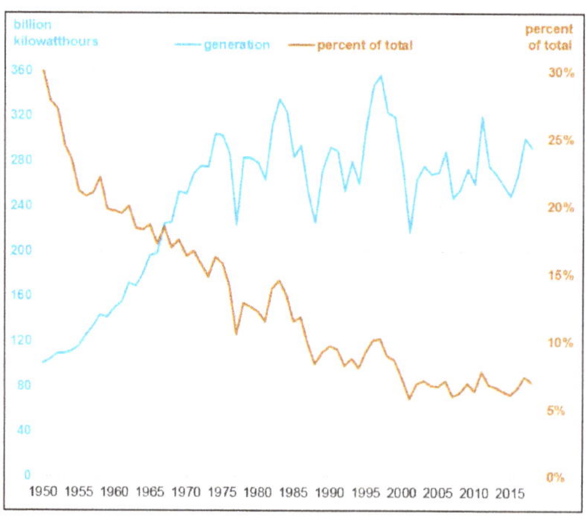

Hydroelectric Power is Produced with Moving Water

Because the source of hydroelectric power is water, hydroelectric power plants are usually located on or near a water source. The volume of the water flow and the change in elevation (or fall) from one point to another determine the amount of available energy in moving water. Swiftly flowing water in a big river, such as the Columbia River that forms the border between Oregon and Washington, carries a great deal of energy in its flow. Water descending rapidly from a high point, such as Niagara Falls in New York, also has substantial energy in its flow.

At both Niagara Falls and the Columbia River, water flows through a pipe, or penstock, then pushes against and turns blades in a turbine to spin a generator to produce electricity. In a run-of-the-river system, the force of the current applies pressure on a turbine. In a storage system, water accumulates in reservoirs created by dams and is released as needed to generate electricity.

Hydropower Advantages and Disadvantages

Hydropower has several advantages. Once a dam has been built and the equipment installed, the energy source—flowing water—is free. It's a clean fuel source renewed by snow and rainfall. Hydropower plants can supply large amounts of electricity, and they are relatively easy to adjust for demand by controlling the flow of water through the turbines from businesses looking to develop the region.

The pristine Big Darby Creek in the summer. This stream is under risk.

But big dam projects can disrupt river ecosystems and surrounding communities, harming wildlife and forcing out residents. The Three Gorges Dam, for example, displaced an estimated 1.2 million people and flooded hundreds of villages.

Dams also prevent fish such as salmon from swimming upstream to spawn. While equipment such as fish ladders are designed to help salmon go up and over dams and enter upstream spawning areas, such measures aren't always effective. In some cases, fish are collected and trucked around the obstacles. Still, the presence of hydroelectric dams can often change migration patterns and hurt fish populations. In the Columbia

River Basin in the Pacific Northwest, for example, salmon and steelhead have lost access to about 40 percent of their historic habitat because of dams.

Hydropower plants can also cause low dissolved oxygen levels in the water, which is harmful to river habitats. Other wildlife can be affected as well: In Indonesia, a hydroelectric project threatens rare Tapanuli orangutans because it stands to fragment their habitat.

Climate change and the increased risk of drought are also having an impact on the world's hydropower plants. In the western U.S., carbon dioxide emissions over a 15-year period were 100 megatons higher than they normally would have been, according to a 2018 study, as utilities turned to coal and gas to replace hydropower lost to drought.

Even the promise of carbon-free electricity from hydropower has been undermined by revelations that decaying organic material in reservoirs releases methane, a potent greenhouse gas that contributes to global warming.

However, some argue that the environmental impacts of hydroelectric power can be mitigated and remain low compared with burning fossil fuels. In some places, small hydro projects can take advantage of existing water flows or infrastructure. Special water intakes and turbines can help make sure water released from a dam is better aerated to address the problem of low dissolved oxygen. Dams can be planned more strategically to allow fish passages, for example, while water flows at existing dams can be calibrated to give ecosystems more recovery time from flooding cycles. And research continues on ways to make hydropower projects more friendly to the ecosystems around them.

A growing movement is also working to tear down dams that are no longer functioning or needed around the world, with the aim at restoring more natural rivers and the many benefits they provide to wildlife and people, including recreation.

HYDRAULIC HEAD

The hydraulic head is a value that measures the amount of mechanical energy available in water in a river, stream or even lake. The hydraulic head is equivalent to the water level in a static (non-flowing) water body. In its simplest form hydraulic head is a measurement of the height of a static water column above an arbitrary point, usually expressed in meters (or feet in the US). The higher the water level or hydraulic head, the more energy that the water at a specific location has.

In a hydroelectric facility, the amount of energy harnessed depends on the difference between the headwater level in the reservoir upstream of the dam and the tailwater level below the hydroelectric dam. This is known as the hydraulic head difference and it represents the amount of energy that can be transformed into electricity by the turbines and generators. More detailed calculations show that not only the difference in height of the water determines how much energy can be harnessed, but there are a variety of losses known as head losses. These head losses occur as a result of friction in pipes. When head losses are accounted for, the actual amount of energy that can be harnessed is reduced. This reduced value for hydraulic head with the losses accounted for is known as the effective head.

The water in the reservoir, which is at a higher altitude has more gravitational potential energy because it is at a higher vertical position than the tail race. The energy used to generate electricity by moving the turbines arises from the utilization of gravitational potential energy of the reservoir water as it moves down through the penstocks. The use of Bernoulli's equation can help illustrate this process. Overall, hydraulic head is a way to represent the energy of energy of stored a fluid - in this case water - per unit weight.

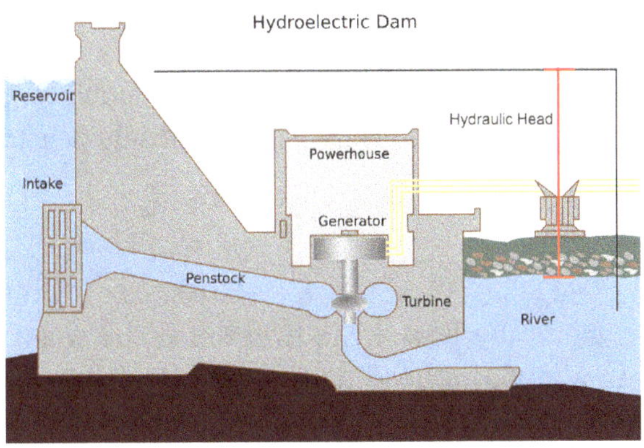

A labelled diagram of a hydroelectric plant, including the hydraulic head.

The value for hydraulic head is used in the hydroelectric power equation to determine the available power of a hydroelectric system. It is one of the main components in this equation, represented by:

$$P = \rho Q g \Delta h$$

where,

- P is the amount of power being calculated, measured in joules per second (J/s) which is also known as a watt (W).
- Δh is the hydraulic head difference across the dam or turbine, expressed in meters (m).
- ρ is the density of the fluid, measured in kilograms per unit volume (kg/m^3).
- Q is the volumetric hydroelectric discharge (the flow rate) of the fluid, measured in volume per unit time (m^3/s).
- g is acceleration due to gravity, measured in meters per second (m/s).

Note that the larger the head difference in the expression above, the more potential mechanical energy the water in the reservoir has.

Classifications

There are three main classifications of dames based on the hydraulic head differences: high, medium, and low. The characteristics of these types are described below:

High Head

Head differences of 100 meters or higher are considered high head. In this type of plant, water travelling through the turbine comes from a significantly higher elevation, meaning that the system needs a smaller volume of water to produce an equivalent amount of energy. These systems generally also require smaller turbines since there is less water flowing through the turbine.

In addition to a smaller turbine, a longer penstock is needed to guide the water down from the high elevation reservoir. Generally, large hydroelectric installations are either high or medium head.

Medium Head

Medium head systems generally have head differences between between 10 and 100 meters. The penstock in a medium head dam is slightly shorter than a high head dam as there is less of an elevation drop. This type of dam relies both on a significant volume of water flowing and a significant drop in height of the water.

Low Head

Low head dams are usually classified as systems with head differences of around 10 meters or less. These low head hydro turbines are generally used in facilities such as run-of-the-river systems where there is a flowing river with little elevation change.

These low head systems usually transport large volumes of water and thus require larger turbines to efficiently convert water energy into electricity. In these installations, a large dam isn't needed to barricade water as there is very little water storage.

Head Losses

The height difference between the water reservoir and the tail race in a hydroelectric facility is termed the gross head and it determines the maximum amount of power that can be produced. However, energy losses occur whenever water flows through pipes due to friction and this reduces the actual rate of power production. These losses reduce the overall head and result in a number known as the effective head or net head. Effective head is equal to the gross head minus the sum of all head losses. Head losses occur in all hydroelectric facilities, and are classified as major and minor head losses. These hydraulic head losses are measured, calculated, and expressed in the same way hydraulic head is, in the equivalent height of water in meters. In this case, you subtract the amount of power lost through head losses from the overall gross power to get the actual net power one can obtain. The equation can be shown as:

$$P_{net} = P_{gross} - P_{loss}$$

Types of head losses include:

- Major head losses: Major head losses arise mainly from friction in the pipes, and occur over long lengths of pipe, such as in the penstock.
- Minor head losses: Minor head losses arise from any other place besides friction in the pipes. Essentially, any place that the pipe bends or the velocity of the water changes, minor losses exist.

Despite the names, major head losses are not always greater than minor head losses - it is simply the combination of both minor and major head losses that are important in determining what the net power of the hydroelectric power plant is.

LOW HEAD HYDROPOWER

Low head hydropower applications use tidal flows or rivers with a head of 20 metres (66 ft) or less to produce energy. These applications may not need to dam or retain water to create hydraulic head. Using the drop in a river or tidal flows to create electricity may provide a renewable energy source that will have a minimal impact on the environment.

Comparison to Conventional Hydro

Most current hydroelectric projects use a large hydraulic head to power turbines to generate electricity. The hydraulic head either occurs naturally, such as a waterfall, or is created by constructing a dam in a river valley, creating a reservoir. Using a controlled release of water from the reservoir drives the turbines. The costs and environmental impacts of constructing a dam can make traditional hydroelectric projects unpopular in some countries.

Damless hydro captures the kinetic energy of rivers, channels, spillways, irrigation systems, tides and oceans without the use of dams.

Construction of a dam and reservoir may have harmful environmental effects. For example, the damming of a river may “block the movement both of fish upstream to spawn and of silt downstream to fertilize fields". Where sites aren't cleared "the vegetation overwhelmed by the rising water decays to form methane – a far worse greenhouse gas than carbon dioxide", particularly in the tropics.

Since no dam is required, low-head hydro may dramatically reduce the following:

- The safety risks (of having a dam), avoiding the risk of a flash flood caused by a breached dam.
- Environmental and ecological complications:
    - Need for fish ladders.
- Regulatory issues.
- The initial cost of dam engineering and construction.
- Maintenance:
    - Removing Silt accumulation.

However, low-head units are necessarily much smaller in capacity than conventional large hydro turbines, requiring many more to be built for a given annual energy production, with some of the costs of small turbine/generator units being offset by lower civil construction costs. Just as for large hydro, not every site can be economically and ecologically developed; sites may be too far from customers to be worth installation of a transmission line, or may lie in areas particularly sensitive for wildlife.

Another potentially promising type of low head hydro power is dynamic tidal power, a novel and unapplied method to extract power from tidal movements. Although a dam-like structure is required, no area is enclosed, and therefore most of the benefits of 'damless hydro' are retained, while providing for vast amounts of power generation.

Hydrokinetic Turbines

A "Hydrokinetic" turbine is an integrated turbine generator to produce electricity in a free flow environment. It does not need a dam or diversion. Instream Energy Systems has coined the phrase *Instream Energy Generation Technology* or IEGT places turbines in rivers, man made channels, tidal waters, or ocean currents. These turbines use the flow of water to turn them, thus generating electricity for the power grid on nearby land. In effect, IEGT is like planting windmills in the water and is environmentally friendly. While hydrokinetic includes generation from ocean tides, currents and waves, many researchers believe its most practical application in the near term is likely to be in rivers and streams.

Instream Energy Systems has deployed a 25 kilowatt hydrokinetic turbine in a canal just outside Yakima Washington. They have the ability to deploy multiple turbines along a river, canal, or aqueduct in order to maximize their energy harvest with minimal impact on the existing canals. Two major benefits of hydrokinetic deployment are the reliability and the predictability of the water delivery systems. Unlike wind, the water deliveries are planned well in advance and therefore utilities can accurately predict when more power will be coming online. A 35 kilowatt hydrokinetic turbine has been installed in the Mississippi River near Hastings, Minnesota. Underwater tidal turbines are propelled by tidal currents. If the viable river and estuary turbine locations are made into hydroelectric power sites "researchers estimate that the United States rivers and estuaries could provide up to 130,000 gigawatt-hours per year – about half the yearly production of the country's dams."

Types of Low Head Turbines

Turbines suitable for use in very low head applications are different from the Francis, propeller, Kaplan,or Pelton types used in more conventional large hydro.

Different types of low head application turbines are:

- Axial Flow Rotor Turbine: This type of turbine consists of a concentric hub with radial blades, resembling a wind mill. Either a built in electrical generator or a hydraulic pump which turns an electrical generator on land provides the electricity.
- Open Center Fan Turbine: These turbines consists of two donut shaped turbines which rotate in the opposite direction of the current. This in turn runs a hydraulic pump that in turn drives a standard electrical generator.
- Helical Turbine: This type of turbine has hydrofoil sections that keep the turbine oriented to the flow of the water. The leader edge of the blades turns in the direction of the water.
- Cycloidic Turbine: The cycloidic turbine resembles a paddle wheel, where the flow of the water turns the wheel with lift and drag being optimized. Lift or flutter vanes looks like a huge Venetian blind.
- Hydroplane blades: They are made to oscillate by the flowing water, thus generating electricity.
- FFP Turbine Generator: This type of turbine uses a rim-mounted, permanent magnet, direct-drive generator with front and rear diffusers and one moving part (the rotor) to maximize efficiency.
- Gravitation Water Vortex Power Plant: This type of hydro power plant use the power of a gravitation water vortex, which only exists at low head.

Installation of Turbines in River Current

The turbines can be installed in a variety of ways, multiple banks set on pilings driven into the river beds or mounted on existing river structures such as bridge piers. These turbines operate in a "free flow" environment that does not require the damming or diversion of rivers. This approach does not disrupt natural ecosystems or interfere with aquatic and marine life. The turbine generators can be attached to bridge abutments or pilings, which minimizes disruption to river beds.

Turbines are to be deployed in arrays of multiple units spaced no less than 15 metres (49 ft) apart where the site conditions, depth, and needed infrastructure are suitable. Exact depth and spacing is determined based on site conditions, including current flows and water depth. Since the turbines do not block waterways, and the water passing through the device is not subject to high pressure, these systems are designed to not impede or damage fish or other wildlife. This entire structure is installed at a depth that avoids any interference with recreational or navigational uses of a water resource. The power will be transmitted by cable to conversion equipment located on shore. The conversion equipment will convert the power from DC to AC, adjust the voltage and connect to the power grid. This approach will be better suited to river current where the flow is in one direction as opposed to ocean shore locations.

A competing idea is to suspend the turbines from a floating barge. The turbines suspended from the bottom of a floating barge can accommodate changes in flow. The barges can be deployed and have the generators come on line more quickly with fewer disturbances to the river bed. The obvious disadvantage to the barge system would be interference with navigation and recreational use of the waterway. This system does have some advantages, installation costs may be less depending on river bottom conditions and maintenance and repair would be somewhat easier. Concern over the impact on seasonal flooding and ice conditions also must be considered with the barge system.

Tidal Power

Tidal flow occurs due to the moving mass of water with speed and direction as caused by the gravitational forces of the sun and the moon, and centrifugal and inertial forces on the Earth's waters. Due to its proximity to the earth, the moon exerts roughly twice the tide raising force of the sun. The gravitational forces of the sun and moon and the centrifugal/inertial forces caused by the rotation of the earth around the center of mass of the earth-moon system create two "bulges" in the Earth's oceans: one closest to the moon, and the other on the opposite side of the globe. This kind of energy is unique and different from traditional hydropower that has been around for centuries. There is no need to build a dam. Essentially a turbine is stuck in naturally flowing water. As the water flows, it turns a turbine. That is converted to electricity.

Tidal basin locations can also be developed using the low flow turbine technology as well. These areas are limited to ocean side locations and the difficulty associated with rotating the turbines to adjust to the direction of the tidal flow must also be accounted for. It would appear that the turbines suspended from under a floating barge would be better suited to the tidal application. The barge itself can be turned to face the direction of the tidal flow. It may also be more difficult to provide the areas for power conversion and connection to the power grid given the limited areas that can be developed to utilize tidal flows. Several demonstration projects are underway to study the feasibility of the tidal basin locations. Tidal turbines are a new technology used for tidal energy. They are similar to wind turbines and are arranged underwater in rows. They work best in areas with strong tides. They are also the least environmentally damaging of all the tidal power technologies, since they do not interfere with migration paths and the impact on basin bed is less as no construction is needed in the waterway itself.

In order for tidal power to work successfully it requires a tide difference of at least 5 metres (16 ft). Unfortunately there are only a few places where this occurs. This means tidal power plants cannot just be constructed anywhere. There are only a handful of sites on Earth with this type of tidal range. A demonstration project has begun in New York City. In the last four years, the federal commission has approved nearly a dozen permits to study tidal sites. Applications for about 40 others, all filed in 2006, are under review. No one has applied for a development license, Miller said. The site that is furthest along in testing lies in New York's East River, between the boroughs of Manhattan and Queens, where Verdant Power plans to install two underwater turbines this month as part of a small pilot project.

Ocean and tidal currents can provide an indefinite supply of emission-free renewable energy. Since tidal and river currents exist everywhere in the world and are either constantly flowing or extremely predictable, converting the energy in these currents to electricity could provide a predictable, reliable and, in some cases, base load supply of electricity to the electric power systems or remote sites in many parts of the world. 70% of the world's population lives within 320 kilometres (200 mi) of an ocean. Accordingly, ocean current energy could become a vital part of the world's energy future.

Environmental Impact of Low Head Hydropower

A number of concerns have been raised about the environmental impacts of river current and tidal devices. Among the most important of these are:

- Marine life: Concerns have been raised about the danger of rotating blades to marine animals, such as seals and fish, from wave and tidal devices. There is no evidence that this is a significant problem. Such devices may actually benefit the local fauna by creating non-fishing 'havens' and structures such as anchoring devices may create new reefs for fish colonization.
- Sea bed: By altering wave patterns and tidal streams, devices will undoubtedly have an effect, for example, on the deposition of sediment. Research carried out to date would seem to indicate that the effects would not be significant, and may even be positive, for example by helping to slow down coastal erosion. (This is particularly pertinent in light of evidence that waves have steadily increased

in size in the recent past.) The sea in the lee of devices would almost certainly be calmer than normal, but, it has been suggested, this would help in creating more areas for activities such as water sports or yachting.

- Landscape: Most wave and tidal energy devices would be invisible from the shore. They would have none of the problems of visual and noise pollution that older versions of wind turbines engender. The main impact would probably be from the extensive transmission lines needed to take the energy from the shoreline to final users. This problem would have to be addressed, possibly by using underground transmission lines.
- Fishing and shipping activities: Offshore wave and tidal devices would almost certainly require areas to be closed to fishing and shipping activities. The siting of such devices would have to be negotiated, therefore, with relevant local groups (for example, fishermen), as well as with national and international bodies.

Implementation and Regulations

Government Regulation

Most government regulation comes from the use of waterways. Most low head water turbine systems are smaller engineering projects than traditional water turbines. Even so, one needs to obtain permission from state and federal government institutions before implementing these systems. Some of the constraints faced with these systems in larger waterways are making sure waterways can still be used for boats and making sure that routes of migration of fish are not disturbed.

Government Subsidies

US government subsidies can be obtained for implementation of small-scale hydro facilities most easily through federal grants, namely green energy grants. a specific example is the Renewable Electricity Production Tax Credit. This is a federal tax credit aimed at promoting renewable energy resources. To qualify, the hydro source must have a minimum capacity of 150 kW. This subsidy is given for the first ten years of production. Organizations receive $.011/kWh.

Public Perception

Since these are sustainable energy source, are non detrimental to the water sources they utilize and are visually not an eyesore, they are well regarded within the public sphere. However, there is little public and industrial knowledge of these systems as they are still being tested to “answer real-world questions”. As such, proponents and manufacturers of these systems have tried to bring them into public knowledge.

TIDAL POWER

Tidal Energy or Tidal Power as it is also called, is another form of hydro power that utilises large amounts of energy within the oceans tides to generate electricity. Tidal Energy is an "alternative energy" that can also be classed as a "renewable energy source", as the Earth uses the gravitational forces of both the moon and the sun everyday to move vast quantities of water around the oceans and seas producing tides.

As the Earth, its Moon and the Sun rotate around each other in space, the gravitational movement of the moon and the sun with respect to the earth, causes millions of gallons of water to flow around the Earth's oceans creating periodic shifts in these moving bodies of water. These vertical shifts of water are called "tides".

When the earth and the moons gravity lines up with each other, the influences of these two gravitational forces becomes very strong and causes millions of gallons of water to move or flow towards the shore creating a "high tide" condition. Likewise when the earth and the moons gravity are at 900 to each other, the influences of these two gravitational forces is weaker and the water flows away from the shore as the mass of water moves to another location on the earth, creating a "low tide" condition. This ebbing and flowing of the tides happens twice during each period of rotation of the earth with stronger weekly and annual lunar cycles superimposed onto these tides.

When the moon is in perfect alignment with the earth and the sun, the gravitational pull of the moon and sun together becomes much stronger than normal with the high tides becoming very high and the low tides becoming very low during each tidal cycle. Such tides are known as spring tides (maximum). These spring tides occur during the full or new moon phase.

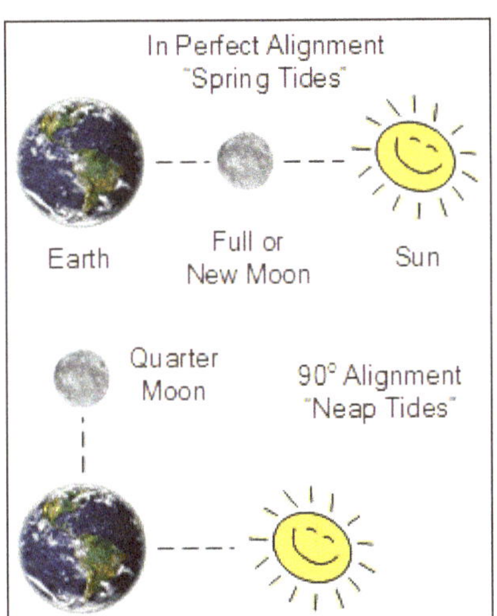

The other tidal situation arises during neap tides (minimum) when the gravitational pull of the moon and the sun are against each other, thus cancelling their effects. The net result is a smaller pulling action on the sea water creating much smaller differences between the high and low tides thereby producing very weak tides. Neap tides occur during the quarter moon phase. Then spring tides and neap tides produce different amounts of potential energy in the movement of the sea water as their effects differ from the regular high and low sea levels and we can use these tidal changes to produce renewable energy. So we can say that the tides are turning for alternative energy.

So we now know that the constant rotational movement of the earth and the moon with regards to each other causes huge amounts of water to move around the earth as the tides go in and out. These tides are predictable and regular resulting in two high tides and two low tides each day with the level of the oceans constantly moving between a high tide and a low tide, and then back to a high tide again. The time taken for a tidal cycle to happen is about 12 hours and 24 minutes (called the "diurnal cycle") between two consecutive high tides allowing Oceanographers and Meteorologist to accurately predict the ebb and flow of the tides around the oceans many years in advance.

The main big advantage of this is that the tides are therefore perfectly predictable and regular unlike wind energy or solar energy, allowing miles of coastline to be used for tidal energy exploitation and the larger the tidal influence, the greater the movement of the tidal water and therefore the more potential energy that can be harvested for power generation. Therefore Tidal Energy can be considered as a renewable energy source as the oceans energy is replenished by the sun as well as through tidal influences of the moon and suns gravitational forces.

Tidal Energy Generation

Since the position of the earth and the moon with respect to the sun changes throughout the year, we can utilise the potential energy of the water contained in the daily movement of the rising and falling sea levels to generate electricity. The generation of electricity from tides is similar in many ways to hydro-electric generation we looked at in the hydro energy tutorials. The difference this time is that the water flows in and out of the turbines in both directions instead of in just one forward direction.

Tidal energy, just like hydro energy transforms water in motion into a clean energy. The motion of the tidal water, driven by the pull of gravity, contains large amounts of kinetic energy in the form of strong tidal currents called tidal streams. The daily ebbing and flowing, back and forth of the oceans tides along a coastline and into and out of small inlets, bays or coastal basins, is little different to the water flowing down a river or stream.

The movement of the sea water is harnessed in a similar way using waterwheels and turbines to that used to generate hydro electricity. But because the sea water can flow in both directions in a tidal energy system, it can generate power when the water is flowing in and also when it is ebbing out. Therefore, tidal generators are designed to produce power when the rotor blades are turning in either direction. However, the cost of reversible electrical generators are more expensive than single direction generators.

Different Types of Tidal Energy Systems

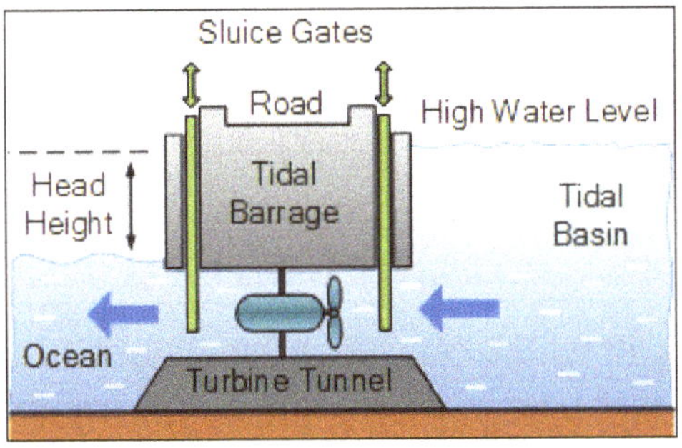

Tidal Barrage – A Tidal Barrage is a type of tidal power generation that involves the construction of a fairly low dam wall, known as a "barrage" and hence its name, across the entrance of a tidal inlet or basin creating a tidal reservoir. This dam has a number of underwater tunnels cut into its width allowing sea water to flow through them in a controllable way using "sluice gates". Fixed within the tunnels are huge water turbine generators that spin as the water rushes past them generating tidal electricity.

Tidal barrages generate electricity using the difference in the vertical height between the incoming high tides and the outgoing low tides. As the tide ebbs and flows, sea water is allowed to flow in or out of the reservoir through a one way underwater tunnel system. This flow of tidal water back and forth causes the water turbine generators located within the tunnels to rotate producing tidal energy with special generators used to produce electricity on both the incoming and the outgoing tides.

The one disadvantage of Tidal Barrage Generation, is that it can only generate electricity when the tide is actually flowing either "in" or "out" as during high and low tide times the tidal water is stationary. However, because tides are totally predictable, other power stations can compensate for this stationary period when there is no tidal energy being produced. Another disadvantage of a tidal barrage system, is the environmental and ecological effects that a long concrete dam may have on the estuaries they span.

Tidal Stream – A Tidal Stream Generation system reduces some of the environmental effects of tidal barrages by using turbine generators beneath the surface of the water.

Major tidal flows and ocean currents, like the Gulf Stream, can be exploited to extract its tidal energy using underwater rotors and turbines.

Tidal stream generation is very similar in principal to wind power generation, except this time water currents flow across a turbines rotor blades which rotates the turbine, much like how wind currents turn the blades for wind power turbines. In fact, tidal stream generation areas on the sea bed can look just like underwater wind farms.

Unlike off-shore wind power which can suffer from storms or heavy sea damage, tidal stream turbines operate just below the sea surface or are fixed to the sea bed. Tidal streams are formed by the horizontal fast flowing volumes of water caused by the ebb and flow of the tide as the profile of the sea bed causes the water to speed up as it approaches the shoreline.

As water is much more denser than air and has a much slower flow rate, tidal stream turbines have much smaller diameters and higher tip speed rates compared to an equivalent wind turbine. Tidal stream turbines generate tidal power on both the ebb and flow of the tide. One of the disadvantages of Tidal Stream Generation is that as the turbines are submerged under the surface of the water they can create hazards to navigation and shipping.

Other forms of tidal energy include tidal fences which use individual vertical-axis turbines that are mounted within a fence structure, known as the caisson, which completely blocks a channel and force water through them. Another alternative way of harnessing tidal power is by using an "oscillating tidal turbine". This is basically a fixed wing called a Hydroplane positioned on the sea bed. The hydroplane uses the energy of the tidal stream flowing past it to oscillate its giant wing, similar to a whales flipper, up and down with the movement of the tidal currents. This motion is then used to generate electricity. The angle of the hydroplane to the flow of the tide can be varied to increase efficiency.

Tidal energy is another form of low-head hydro power that is completely carbon neutral like wind and hydro energy. Tidal power has many advantages compared to other forms of renewable energy with its main advantage being that it is predictable. However, like many other forms of renewable energy, tidal energy also has its disadvantages such as its inflexible generation times dependant upon the tides and the fact that it operates in the hostile conditions of the oceans and seas. So here are some of the advantages and disadvantages associated with "tidal energy".

Advantages of Tidal Energy

- Tidal energy is a renewable energy resource because the energy it produces is free and clean as no fuel is needed and no waste bi-products are produced.
- Tidal energy has the potential to produce a great deal of free and green energy.
- Tidal energy is not expensive to operate and maintain compared to other forms of renewable energies.

- Low visual impact as the tidal turbines are mainly if not totally submerged beneath the water.
- Low noise pollution as any sound generated is transmitted through the water.
- High predictability as high and low tides can be predicted years in advance, unlike wind.
- Tidal barrages provide protection against flooding and land damage.
- Large tidal reservoirs have multiple uses and can create recreational lakes and areas where before there were none.

Disadvantages of Tidal Energy

- Tidal energy is not always a constant energy source as it depends on the strength and flow of the tides which themselves are effected by the gravitational effects of the moon and the sun.
- Tidal Energy requires a suitable site, where the tides and tidal streams are consistently strong.
- Must be able to withstand forces of nature resulting in high capital, construction and maintenance costs.
- High power distribution costs to send the generated power from the submerged devices to the land using long underwater cables.
- Intermittent power generation, only generates power ten hours a day during the ebb and flow of the tides.
- Changes to estuary ecosystem and an increase in coastal erosion where the tides are concentrated.
- Build up of silt, sediments and pollutants within the tidal barrage from rivers and streams flowing into basin as it is unable to flow out into the sea.
- Danger to fish and other sea-life as they get stuck in the barrage or sucked through the tidal turbine blades.

References

- Hydropower: irena.org, Retrieved 29 March, 2019
- Hydropower-home, energyexplained: eia.gov, Retrieved 16 January, 2019
- Hydropower, global-warming, environment: nationalgeographic.com, Retrieved 30 August, 2019
- Hydraulic-head, encyclopedia: energyeducation.ca, Retrieved 15 July, 2019
- Tidal-energy: alternative-energy-tutorials.com, Retrieved 30 April, 2019

2

Hydroelectricity and its Type

The electricity produced by generators which are pushed by the movement of water is termed as hydroelectricity. It can be produced at various scales which are generally classified as small hydro, micro hydro and pico hydro. These types of hydroelectricity have been thoroughly discussed in this chapter.

HYDROELECTRICITY

Hydroelectricity is electricity produced by hydropower—that is, the energy of moving water. It is the world's leading form of renewable energy. In 2005, it accounted for over 63 percent of the total renewable energy. That same year, hydroelectricity supplied about 715,000 megawatts (or 19 percent) of the world's electricity (compared to 16 percent in 2003). Although large hydroelectric installations generate most of the world's hydroelectricity, small hydro schemes are particularly popular in China, which has over 50 percent of the world's small hydro capacity.

The Nagarjuna dam & hydroelectric plant, India.

On the downside, hydroelectric projects can disrupt ecosystems by reducing access to salmon spawning grounds and damaging habitats of birds. These projects also lead to changes in the downstream river environment, by the scouring of river beds and loss of riverbanks. Large-scale hydroelectric dams, such as the Aswan Dam and the Three Gorges Dam, have created environmental problems both upstream and downstream.

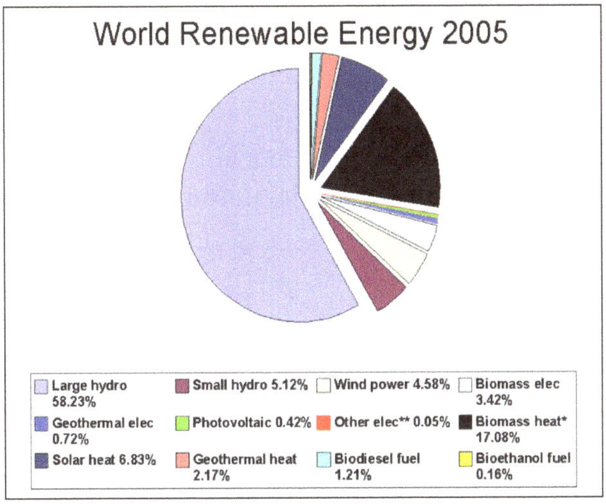

Hydroelectricity is the world's leading renewable energy source.

Electricity Generation

Most hydroelectric power comes from the potential energy of dammed water driving a water turbine and generator. In this case, the energy extracted from the water depends on the volume and on the difference in height between the source and the water's outflow. This height difference is called the head. The amount of potential energy in water is proportional to the head. To obtain very high head, water for a hydraulic turbine may be run through a large pipe called a penstock.

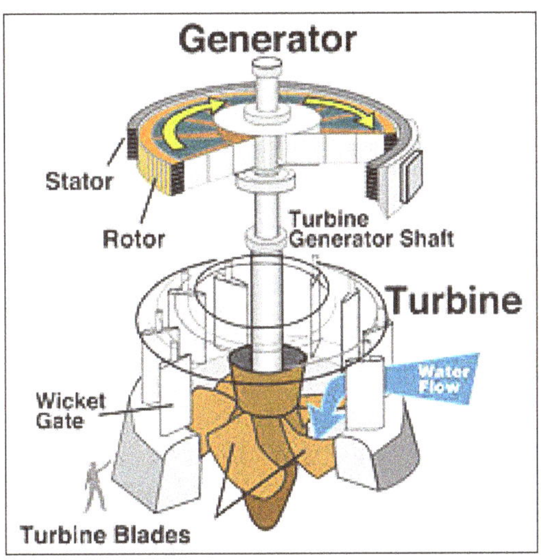

Hydraulic turbine and electrical generator.

Pumped storage hydroelectricity produces electricity to supply high peak demands by moving water between reservoirs at different elevations. At times of low electrical demand, excess generation capacity is used to pump water into the higher reservoir. When there is higher demand, water is released back into the lower reservoir through a turbine. Pumped storage schemes currently provide the only commercially important means of grid energy storage and improve the daily load factor of the generation system.

Less common types of hydro schemes use water's kinetic energyor undammed sources such as run-of-the-river, waterwheels, and tidal power.

Industrial Hydroelectric Plants

Many hydroelectric projects supply public electricity networks, but some are created to serve specific industrial enterprises. Dedicated hydroelectric projects are often built to provide the substantial amounts of electricity needed for aluminium electrolytic plants, for example. In the Scottish Highlands there are examples at Kinlochleven and Lochaber, constructed during the early years of the twentieth century. In Suriname, the "van Blommestein" lake, dam and power station were constructed to provide electricity for the Alcoa aluminium industry. As of 2007 the Kárahnjúkar Hydropower Project in Iceland remains controversial.

Advantages

The upper reservoir and dam of the Ffestiniog pumped storage scheme. 360 megawatts of electricity can be generated within 60 seconds of the need arising.

Economics

The main advantage of hydroelectricity is that it is nearly independent of increases in the cost of fossil fuels such as oil, natural gas, or coal. Fuel is not required, and so it need not be imported. Hydroelectric plants tend to have longer economic lives than fuel-fired generation, with some plants now in service having been built 50 to 100 years ago. Operating labor cost is usually low since plants are automated and have few personnel on site during normal operation.

Where a dam serves multiple purposes, a hydroelectric plant may be added with relatively low construction cost, providing a useful revenue stream to offset the costs of dam operation. It has been calculated that the sale of electricity from the Three Gorges Dam will cover the construction costs after 5 to 7 years of full generation.

Related Activities

Reservoirs created by hydroelectric schemes often provide facilities for water sports, and become tourist attractions in themselves. In some countries, farming fish in the reservoirs is common. Multi-use dams installed for irrigation can support the fish farm with relatively constant water supply. Large hydro dams can control floods, which would otherwise affect people living downstream of the project. When dams create large reservoirs and eliminate rapids, boats may be used to improve transportation.

Greenhouse Gas Emissions

Since no fossil fuel is consumed, emission of carbon dioxide (a greenhouse gas) from burning fuel is eliminated. While some carbon dioxide is produced during manufacture and construction of the project, this is a tiny fraction of the operating emissions of equivalent fossil-fuel electricity generation. However, there may be other sources of emissions.

Disadvantages

Recreational users must exercise extreme care when near hydroelectric dams, power plant intakes and spillways.

Environmental Damage

Hydroelectric projects can be disruptive to surrounding aquatic ecosystems. For instance, studies have shown that dams along the Atlantic and Pacific coasts of North America have reduced salmon populations by preventing access to spawning grounds upstream, even though most dams in salmon habitat have fish ladders installed. Salmon spawn are also harmed on their migration to sea when they must pass through turbines. This has led to some areas transporting smolt downstream by barge during parts of the year. Turbine and power-plant designs that are easier on aquatic life are an active area of research.

Generation of hydroelectric power changes the downstream river environment. Water exiting a turbine usually contains very little suspended sediment, which can lead to scouring of river beds and loss of riverbanks. Since turbines are often opened intermittently, rapid or even daily fluctuations in river flow are observed. For example, in the Grand Canyon, the daily cyclic flow variation caused by Glen Canyon Dam was found to be contributing to erosion of sand bars. Dissolved oxygen content of the water may change from pre-construction conditions. Water exiting from turbines is typically much colder than the pre-dam water, which can change aquatic faunal populations, including endangered species. Some hydroelectric projects also utilize canals, typically to divert a river at a shallower gradient to increase the head of the scheme. In some cases, the entire river may be diverted, leaving a dry riverbed. Examples include the Tekapo and Pukaki Rivers. Large-scale hydroelectric dams, such as the Aswan Dam and the Three Gorges Dam, have created environmental problems both upstream and downstream.

A further concern is the impact of major schemes on birds. Since damming and redirecting the waters of the Platte River in Nebraska for agricultural and energy use, many native and migratory birds such as the Piping Plover and Sandhill Crane have become increasingly endangered.

Greenhouse Gas Emissions

The reservoirs of hydroelectric power plants in tropical regions may produce substantial amounts of methane and carbon dioxide. This is due to plant material in flooded areas decaying in an anaerobic environment, and forming methane, a very potent greenhouse gas. According to the World Commission on Dams report, where the reservoir is large compared to the generating capacity (less than 100 watts per square meter of surface area) and no clearing of the forests in the area was undertaken prior to impoundment of the reservoir, greenhouse gas emissions from the reservoir may be higher than those of a conventional oil-fired thermal generation plant.

In boreal reservoirs of Canada and Northern Europe, however, greenhouse gas emissions are typically only 2 to 8 percent of any kind of conventional thermal generation. A new class of underwater logging operation that targets drowned forests can mitigate the effect of forest decay.

Population Relocation

Another disadvantage of hydroelectric dams is the need to relocate the people living where the reservoirs are planned. In many cases, no amount of compensation can replace ancestral and cultural attachments to places that have spiritual value to the displaced population. Additionally, historically and culturally important sites can be flooded and lost. Such problems have arisen at the Three Gorges Dam project in China, the Clyde Dam in New Zealand, and the Ilısu Dam in Southeastern Turkey.

Dam Failures

Failures of large dams, while rare, are potentially serious—the Banqiao Dam failure in China resulted in the deaths of 171,000 people and left millions homeless, more than some estimates of the death toll from the Chernobyl disaster. Dams may be subject to enemy bombardment during wartime, sabotage, and terrorism. Smaller dams and micro hydro facilities are less vulnerable to these threats.

The creation of a dam in a geologically inappropriate location may cause disasters like the one of the Vajont Dam in Italy, where almost 2000 people died, in 1963.

Comparison with other Methods of Power Generation

Hydroelectricity eliminates the flue gas emissions from fossil fuel combustion, including pollutants such as sulfur dioxide, nitric oxide, carbon monoxide, dust, and mercury in the coal.

Compared to the nuclear power plant, hydroelectricity generates no nuclear waste, nor nuclear leaks. Unlike uranium, hydroelectricity is also a renewable energy source.

Compared to wind farms, hydroelectricity power plants have a more predictable load factor. If the project has a storage reservoir, it can be dispatched to generate power when needed. Hydroelectric plants can be easily regulated to follow variations in power demand.

Unlike fossil-fueled combustion turbines, construction of a hydroelectric plant requires a long lead-time for site studies, hydrological studies, and environmental impact assessment. Hydrological data up to 50 years or more is usually required to determine the best sites and operating regimes for a large hydroelectric plant. Unlike plants operated by fuel, such as fossil or nuclear energy, the number of sites that can be economically developed for hydroelectric production is limited; in many areas the most cost effective sites have already been exploited. New hydro sites tend to be far from population centers and require extensive transmission lines. Hydroelectric generation depends on rainfall in the watershed, and may be significantly reduced in years of low rainfall or snowmelt. Utilities that primarily use hydroelectric power may spend additional capital to build extra capacity to ensure sufficient power is available in low water years.

SMALL HYDRO

Small hydro is the development of hydroelectric generation facilities on a scale corresponding to river discharge and potential, and which is suitable for local community and industry, or to contribute to distributed generation in a regional electricity grid.

The definition of a "small hydro" project varies based on different consideration of what should constitute bottom and top limits from country to country, but most common range, although not binding, in capacity of "small hydro" extend between 1 to 15 megawatts (MW), and can be further subdivide by scale into "mini" (<1MW), "micro" (<100kW), "pico" (<10kW). In contrast many hydroelectric projects are of enormous size, such as the generating plant at the Three Gorges Dam at 22,500 megawatts or the vast multiple projects of the Tennessee Valley Authority. In India, hydro projects up to 25 MW station capacities have been categorized as Small Hydro Power (SHP) projects.

Small hydro projects may be built in isolated areas that would be uneconomic to serve from a national electricity grid, or in areas where a national grid does not exist.

Small power plant of Licq-Athérey (Pyrénées-Atlantiques, France).

An 1895 hydroelectric plant near Telluride, Colorado.

The use of the term "small hydro" varies considerably around the world, the maximum limit is usually somewhere between 10 and 30 MW. While a minimum limit is not usually set, the US National Hydropower Association specifies a minimum limit of 5 MW. In California, hydroelectric generating stations with a maximum capacity of less than 30 MW are classified as small, and are eligible for inclusion in the state's renewable portfolio standard, while hydroelectric generating stations with a higher capacity are classified as large and are not considered renewable. The "small hydro" description may be stretched up to 50 MW in the United States, Canada and China. Small hydro can be further subdivided into mini hydro, usually defined as 100 to 1,000 kilowatts (kW), and micro hydro which is 5 to 100 kW. Micro hydro is usually the application of hydroelectric power sized for smaller communities, single families or small enterprise. The smallest installations are pico hydro, below 5 kW.

Since small hydro projects usually have correspondingly small civil construction work and little or no reservoir, they are seen as having a relatively low environmental impact compared to large hydro.

Growth

According to a report by REN21, during 2008 small hydro installations grew by 28% over year 2005 to raise the total world small hydro capacity to 85 gigawatts (GW). Over 70% of this was in China (with 65 GW), followed by Japan (3.5 GW), the United States (3 GW) and India (2 GW). Global growth is expected to be 2.8% annually until the mid-2020s when capacity will be about 150 gigawatts.

China planned to electrify a further 10,000 villages between 2005 and 2010 under their China Village Electrification Program, including further investments in small hydro and photovoltaics. By 2010, China had 45,000 small hydro installations, especially in rural areas, producing 160 Twh annually. Over 50% of the world's potential small hydro power was found in Asia; however, the report noted, "It is possible in the future that more small hydropower potential might be identified both on the African and American continents".

In the mountains and rain forests of British Columbia, Canada there are a great many sites suitable for hydro development. However environmental concerns towards large reservoirs after the 1980s halted new dam construction. The solution to coping with increased demand was to offer contracts to independent power producers, who have built 100 run of the river projects under 50 MW. Power production without reservoirs varies dramatically, but older conventional dams retain or release water to average out production though the year. In 2014 these independent producers generated 18,000 GWh from 4,500 MW of capacity.

Generation

Historic Ottenbach Small Hydro with original equipment of 1920 in Ottenbach, Switzerland, still running for guided visits.

Hydroelectric power is the generation of electric power from the movement of water. A hydroelectric facility requires a dependable flow of water and a reasonable height for the water to fall, called the head. In a typical installation, water is fed from a reservoir through a pipe into a turbine. The water flowing through the turbine causes a electrical generator to rotate, converting the motion into electrical energy.

Small hydro may developed by constructing new facilities or through re-development of existing dams whose primary purpose is flood control, or irrigation. Old hydro sites may be re-developed, sometimes salvaging substantial investment in the installation such as penstock pipe and turbines, or just re-using the water rights associated with an abandoned site. Either of these cost saving advantages can make the return on investment for a small hydro site well worth the use of existing sites.

Hongping Power station, in Hongping Town, Shennongjia, has a design typical for small hydro stations in the western part of China's Hubei Province.

Project Design

Many companies offer standardized turbine generator packages in the approximate size range of 200 kW to 10 MW. These "water to wire" packages simplify the planning and development of the site since one vendor looks after most of the equipment supply. Because non-recurring engineering costs are minimized and development cost is spread over multiple units, the cost of such package systems is reduced. While synchronous generators capable of isolated plant operation are often used, small hydro plants connected to an electrical grid system can use economical induction generators to further reduce installation cost and simplify control and operation.

Small "run of the river" projects do not have a conventional dam with a reservoir, only a weir to form a headpond for diversion of inlet water to the turbine. Unused water simply flows over the weir and the headpond may only be capable of a single day's storage, not enough for dry summers or frozen winters when generation may come to a halt. A preferred scenario is to have the inlet in an existing lake.

Countries like India and China have policies in favor of small hydro, and the regulatory process allows for building dams and reservoirs. In North America and Europe the regulatory process is too long and expensive to consider having a dam and a reservoir for a small project.

Small hydro projects usually have faster environmental and licensing procedures, and since the equipment is usually in serial production, standardized and simplified, and the civil works construction is also reduced, the projects may be developed very rapidly. The physically smaller size of equipment makes it easier to transport to remote areas without good road or rail access.

One measure of decreased environmental impact with lakes and reservoirs depends on the balance between stream flow and power production. Reducing water diversions helps the river's ecosystem, but reduces the hydro system's return on Investment (ROI). The hydro system design must strike a balance to maintain both the health of the stream and the economics.

List of Small Installations Worldwide

Africa

- Zengamina, a 700 kW plant in Kalene Hill, Mwinilunga District in northwestern Zambia 2008.

Asia

- Meenvallam Small Hydroelectric Project, Palakkad district, Kerala, India.
- Bario Asal & Arur Layun Micro-Hydro Community Project, Kelabit Highlands, Sarawak, Malaysia 2009.
- Cantigas Mini Hydro Electric Power Plant, San Fernando, Romblon, Philippines.
- Pakpattan Hydro Power Project, a 2.82 MW plant on Pakpattan Canal in Pakpattan District, Pakistan 2015.

Europe

- Green Valleys Project, Brecon Beacons National Park, Wales, United Kingdom. Joint winner of £300K in Big Green Challenge 2009.
- St Catherine's, a National Trust site near Windermere, Westmorland, United Kingdom.

North America

- Ames Hydroelectric Generating Plant, Colorado, United States. On the List of IEEE Milestones.

- Childs-Irving Hydroelectric Facilities, Arizona, United States. 1916 Now decommissioned.
- Snoqualmie Falls, Washington, United States 1957.
- Malibu Hydro, British Columbia, Canada 2005.
- Cloudworks Energy is one of many Independent power producers that have built projects in British Columbia, Canada.

MICRO HYDRO

Micro hydro is a type of hydroelectric power that typically produces from 5 kW to 100 kW of electricity using the natural flow of water. Installations below 5 kW are called pico hydro. These installations can provide power to an isolated home or small community, or are sometimes connected to electric power networks, particularly where net metering is offered. There are many of these installations around the world, particularly in developing nations as they can provide an economical source of energy without the purchase of fuel. Micro hydro systems complement solar PV power systems because in many areas, water flow, and thus available hydro power, is highest in the winter when solar energy is at a minimum. Micro hydro is frequently accomplished with a pelton wheel for high head, low flow water supply. The installation is often just a small dammed pool, at the top of a waterfall, with several hundred feet of pipe leading to a small generator housing. In low head sites, generally water wheels and Archimedes screws are used.

Micro hydro in northwest Vietnam.

Construction

Construction details of a microhydro plant are site-specific. Sometimes an existing mill-pond or other artificial reservoir is available and can be adapted for power production. In general, microhydro systems are made up of a number of components. The most important include the intake where water is diverted from the natural stream, river, or perhaps a waterfall. An intake structure such as a catch box is required to screen out floating debris and fish, using a screen or array of bars to keep out large objects. In temperate climates, this structure must resist ice as well. The intake may have a gate to allow the system to be dewatered for inspection and maintenance.

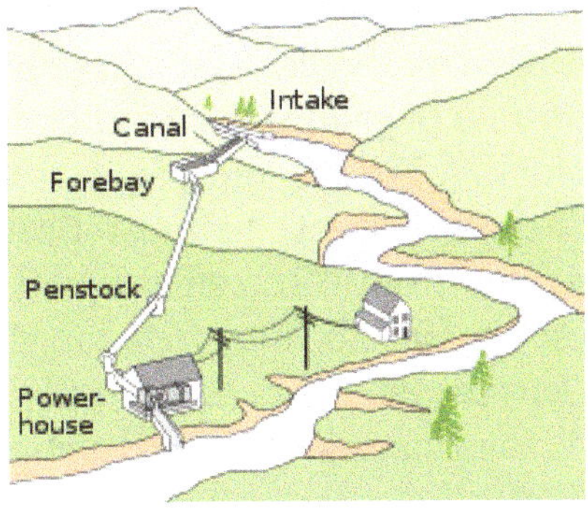

Typical microhydro setup.

The intake is then brought through a canal and then forebay. The forebay is used for sediment holding. At the bottom of the system the water is tunneled through a pipeline (penstock) to the powerhouse building containing a turbine. The penstock builds up pressure from the water that has traveled downwards. In mountainous areas, access to the route of the penstock may provide considerable challenges. If the water source and turbine are far apart, the construction of the penstock may be the largest part of the costs of construction. At the turbine, a controlling valve is installed to regulate the flow and the speed of the turbine. The turbine converts the flow and pressure of the water to mechanical energy; the water emerging from the turbine returns to the natural watercourse along a tailrace channel. The turbine turns a generator, which is then connected to electrical loads; this might be directly connected to the power system of a single building in very small installations, or may be connected to a community distribution system for several homes or buildings.

Usually, microhydro installations do not have a dam and reservoir, like large hydroelectric plants have, relying on a minimal flow of water to be available year-round.

Head and Flow Characteristics

Microhydro systems are typically set up in areas capable of producing up to 100 kilowatts of electricity. This can be enough to power a home or small business facility. This production range is calculated in terms of "head" and "flow". The higher each of these are, the more power available. Hydraulic head is the pressure measurement of water falling in a pipe expressed as a function of the vertical distance the water falls. This change in elevation is usually measured in feet or meters. A drop of at least 2 feet is required or the system may not be feasible. When quantifying head, both gross and net head must be considered. Gross head approximates power accessibility through the vertical distance measurement alone whereas net head subtracts pressure lost due to friction in piping from the gross head. "Flow" is the actual quantity of water falling from a site and is usually measured in gallons per minute, cubic feet per second, or liters per second. Low flow/high head installations in steep terrain have significant pipe costs. A long penstock starts with low pressure pipe at the top and progressively higher pressure pipe closer to the turbine in order to reduce pipe costs.

The available power, in kilowatts, from such a system can be calculated by the equation $P=Q*H/k$, where Q is the flow rate in gallons per minute, H is the static head, and k is a constant of 5,310 gal*ft/min*kW. For instance, for a system with a flow of 500 gallons per minute and a static head of 60 feet, the theoretical maximum power output is 5.65 kW. The system is prevented from 100% efficiency (from obtaining all 5.65 kW) due to the real world, such as: turbine efficiency, friction in pipe, and conversion from potential to kinetic energy. Turbine efficiency is generally between 50-80%, and pipe friction is accounted for using the Hazen–Williams equation.

Regulation and Operation

Typically, an automatic controller operates the turbine inlet valve to maintain constant speed (and frequency) when the load changes on the generator. In a system connected to a grid with multiple sources, the turbine control ensures that power always flows out from the generator to the system. The frequency of the alternating current generated needs to match the local standard utility frequency. In some systems, if the useful load on the generator is not high enough, a load bank may be automatically connected to the generator to dissipate energy not required by the load; while this wastes energy, it may be required if it's not possible to control the water flow through the turbine.

An induction generator always operates at the grid frequency irrespective of its rotation speed; all that is necessary is to ensure that it is driven by the turbine faster than the synchronous speed so that it generates power rather than consuming it. Other types of generator can use a speed control systems for frequency matching.

With the availability of modern power electronics it is often easier to operate the generator at an arbitrary frequency and feed its output through an inverter which produces output at grid frequency. Power electronics now allow the use of permanent magnet alternators that produce wild AC to be stabilised. This approach allows low speed / low head water turbines to be competitive; they can run at the best speed for extraction of energy, and the power frequency is controlled by the electronics instead of the generator.

Very small installations (pico hydro), a few kilowatts or smaller, may generate direct current and charge batteries for peak use times.

Turbine Types

Several types of water turbines can be used in micro hydro installations, selection depending on the head of water, the volume of flow, and such factors as availability of local maintenance and transport of equipment to the site. For hilly regions where a waterfall of 50 meters or more may be available, a Pelton wheel can be used. For low head installations, Francis or propeller-type turbines are used. Very low head installations of only a few meters may use propeller-type turbines in a pit, or water wheels and Archimedes screws. Small micro hydro installations may successfully use industrial centrifugal pumps, run in reverse as prime movers; while the efficiency may not be as high as a purpose-built runner, the relatively low cost makes the projects economically feasible.

In low-head installations, maintenance and mechanism costs can be relatively high. A low-head system moves larger amounts of water, and is more likely to encounter surface debris. For this reason a Banki turbine also called Ossberger turbine, a pressurized self-cleaning crossflow waterwheel, is often preferred for low-head micro hydro systems. Though less efficient, its simpler structure is less expensive than other low-head turbines of the same capacity. Since the water flows in, then out of it, it cleans itself and is less prone to jam with debris.

- Screw turbine (Reverse Archimedes' screw): two low-head schemes in England, Settle Hydro and Torrs Hydro use an Archimedes' screw which is another debris-tolerant design. Efficiency 85%.
- Gorlov: the Gorlov helical turbine free stream or constrained flow with or without a dam.
- Francis and propeller turbines.
- Kaplan turbine: Is a high flow, low head, propeller-type turbine. An alternative to the traditional kaplan turbine is a large diameter, slow turning, permanent magnet, sloped open flow VLH turbine with efficiencies of 90%.
- Water wheel: advanced hydraulic water wheels and hydraulic wheel-part reaction turbine can have hydraulic efficiencies of 67% and 85% respectively. Overshot water wheel maximum efficiency (hydraulic efficiency) is 85%. Undershot water wheels can operate with very low head, but also have efficiencies below 30%.
- Gravitation water vortex power plant: part of the river flow at a weir or natural water fall is diverted into a round basin with a central bottom exit that creates a vortex. A simple rotor (and connected generator) is moved by the kinetic energy. Efficiencies of 83% down to 64% at 1/3 part flow.

Use

Microhydro systems are very flexible and can be deployed in a number of different environments. They are dependent on how much water flow the source (creek, river, stream) has and the velocity of the flow of water. Energy can be stored in battery banks at sites that are far from a facility or used in addition to a system that is directly connected so that in times of high demand there is additional reserve energy available. These systems can be designed to minimize community and environmental impact regularly caused by large dams or other mass hydroelectric generation sites.

Potential for Rural Development

In relation to rural development, the simplicity and low relative cost of micro hydro systems open up new opportunities for some isolated communities in need of electricity. With only a small stream needed, remote areas can access lighting and communications for homes, medical clinics, schools, and other facilities. Microhydro can even run a certain level of machinery supporting small businesses. Regions along the Andes mountains and in Sri Lanka and China already have similar, active programs. One seemingly unexpected use of such systems in some areas is to keep young community members from moving into more urban regions in order to spur economic growth. Also, as the possibility of financial incentives for less carbon intensive processes grows, the future of microhydro systems may become more appealing.

Micro-hydro installations can also provide multiple uses. For instance, micro-hydro projects in rural Asia have incorporated agro-processing facilities such as rice mills – alongside standard electrification – into the project design.

The cost of a micro hydro plant can be between 1,000 and 20,000 U.S. dollars

Advantages

Microhydro power is generated through a process that utilizes the natural flow of water. This power is most commonly converted into electricity. With no direct emissions resulting from this conversion process, there are little to no harmful effects on the environment, if planned well, thus supplying power from a renewable source and in a sustainable manner. Microhydro is considered a "run-of-river" system meaning that water diverted from the stream or river is redirected back into the same watercourse. Adding to the potential economic benefits of microhydro is efficiency, reliability, and cost effectiveness.

Disadvantages

Microhydro systems are limited mainly by characteristics of the site. The most direct limitation comes from small sources with minuscule flow. Likewise, flow can fluctuate seasonally in some areas. Lastly, though perhaps the foremost disadvantage is the distance from the power source to the site in need of energy. This distributional issue as well as the others are key when considering using a microhydro system.

PICO HYDRO

Pico hydro is a term used for hydroelectric power generation of under 5 kW. These generators have proven to be useful in small, remote communities that require only a small amount of electricity – for example, to power one or two fluorescent light bulbs and a TV or radio in 50 or so homes. Even smaller turbines of 200–300 W may power a single home in a developing country with a drop of only one meter. Pico-hydro setups typically are run-of-stream, meaning that a reservoir of water is not created, only a small weir is common, pipes divert some of the flow, drop this down a gradient, and through the turbine before being exhausted back to the stream.

Like other hydroelectric and renewable source power generation, pollution and consumption of fossil fuels is reduced, though there is still typically an environmental cost to the manufacture of the generator and distribution methods.

A pico hydro system made by the Sustainable Vision project from Baylor University.

Small-scale DIY hydroplants

With a growing DIY-community and an increasing interest in environmentally friendly "green energy", some hobbyists have endeavored to build their own hydroelectric plants from old water mills, from kits, or from scratch. The DIY-community has used abandoned water mills to mount a waterwheel and electrical generating components. This approach has also been popularized in the TV-series *It's Not Easy Being Green.* These are usually smaller turbines of ~5 kW or less. Through the internet, the community is now able to obtain plans to construct DIY-water turbines, and there is a growing trend toward building them for domestic requirements. The DIY-hydroelectric plants are now being used both in developed countries and in developing countries, to power residences and small businesses. Two examples of pico hydro power can be found in the towns of Kithamba and Thimba in the Central Province of Kenya. These produce 1.1 kW and 2.2 kW, respectively. Local residents were trained to maintain the hydro schemes. The pico hydro sites in Kenya won Ashden Awards for Sustainable Energy.

Using a pumped-storage system of cisterns and small generators, pico hydro may also be effective for "closed loop" home systems.

Manufacturers

In Vietnam, several Chinese manufacturers have sold pico-powerplants at prices as low as \$20–70 for a powerplant of 300–500 W. However, the devices sold are said to be low in quality and may damage connected equipment if connected improperly.

Sam Redfield of the Appropriate Infrastructure Development Group (AIDG) has developed a pico-hydro generator made from common PVC pipe and a modified Toyota alternator housed in a five gallon bucket. The generator was developed to provide power to communities without access to the electricity grid in developing countries. Envisioned as an energy source to charge cell phones, provide lighting and charge batteries, the generator is designed to be made by artisans with basic skills and can be built for less than \$150.00. The Toyota alternator used in the generator is converted to a permanent magnet alternator allowing it to generate power at low RPMs. The Five Gallon Bucket Hydroelectric Generator was the subject of a work group at the 2008 International Development Design Summit (IDDS) at the Massachusetts Institute of Technology. During the Summer of 2013, an energy project in Abra Malaga, Peru was completed using the bucket generator.

A website has been put together as a forum for ideas and further iterations of the generator that includes a *build manual*. Common centrifugal water pumps, can be operated in reverse to act as turbines. While these machines rarely have optimum hydraulic characteristics when operated as turbines, their availability and low cost makes them attractive.

Transmission Distance

If the power will be used more than 30 meters (100 feet) from the generator, then the transmission distance may be an important consideration. Many small systems use automotive alternators producing 12 VDC, and possibly charging a battery. For example, a 12 V system that produces 1 kW of power has a flow of 80 A and the wire size is 4-gauge. The cost to run two strands of wire 1000' is \$2400(US). To avoid such a large wire cost a higher voltage and lower current is required. If a 240 VAC alternator is used instead the flow is only 4 A over 1000' of 18 gauge wire costing \$180(US). The cost of wire resulted in North America using 120/240 VAC after DC voltage lost the War of Currents in the late 1800s. Another approach to reduce wire costs is to have a 12 VDC alternator with a short high current connection to an inverter outputting 120 VAC or 240 VAC at a much lower current on a long length of thinner wire.

PUMPED-STORAGE HYDROELECTRICITY

Pumped-storage hydroelectricity (PSH), or pumped hydroelectric energy storage (PHES), is a type of hydroelectric energy storage used by electric power systems for load balancing. The method stores energy in the form of gravitational potential energy of water, pumped from a lower elevation reservoir to a higher elevation. Low-cost surplus off-peak electric power is typically used to run the pumps. During periods of high electrical demand, the stored water is released through turbines to produce electric power. Although the losses of the pumping process makes the plant a net consumer of energy overall, the system increases revenue by selling more electricity during periods of *peak demand*, when electricity prices are highest.

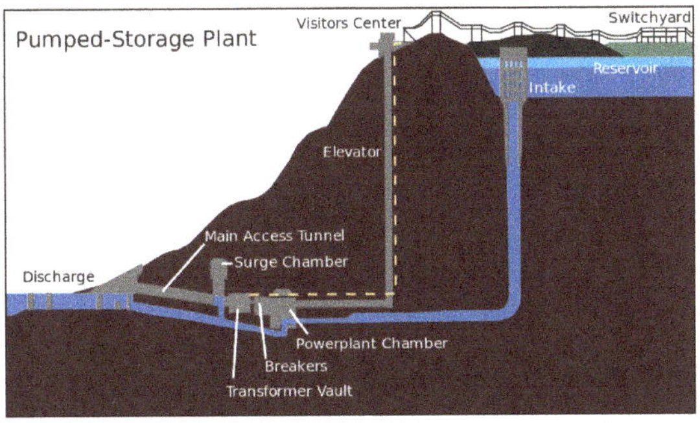

Diagram of the TVA pumped storage facility at Raccoon Mountain Pumped-Storage Plant.

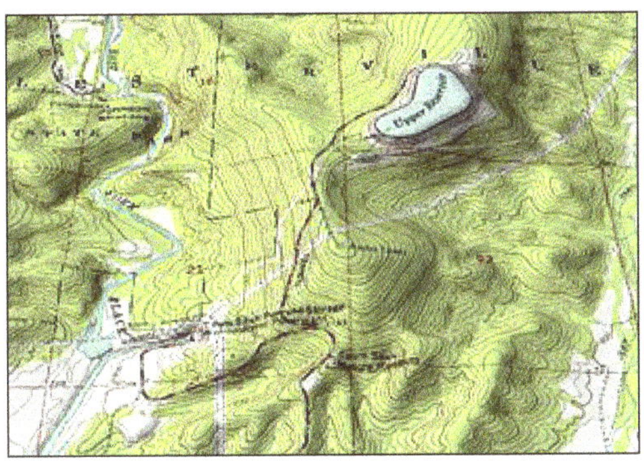

Shaded-relief topo map of the Taum Sauk pumped storage plant in Missouri.

Pumped-storage hydroelectricity allows energy from intermittent sources (such as solar, wind) and other renewables, or excess electricity from continuous base-load sources (such as coal or nuclear) to be saved for periods of higher demand. The reservoirs used with pumped storage are quite small when compared to conventional hydroelectric dams of similar power capacity, and generating periods are often less than half a day.

Pumped storage is the largest-capacity form of grid energy storage available, and, as of 2017, the United States Department of Energy Global Energy Storage Database reports that PSH accounts for over 95% of all active tracked storage installations worldwide, with a total installed nameplate capacity of over 184 GW, of which about 25 GW are in the United States. The *round-trip* energy efficiency of PSH varies between 70%–80%, with some sources claiming up to 87%. The main disadvantage of PSH is the specialist nature of the site required, needing both geographical height and water availability. Suitable sites are therefore likely to be in hilly or mountainous regions, and potentially in areas of outstanding natural beauty, and therefore there are also social and ecological issues to overcome. Many recently proposed projects, at least in the U.S., avoid highly sensitive or scenic areas, and some propose to take advantage of "brownfield" locations such as disused mines.

Basic Principle

At times of low electrical demand, excess generation capacity is used to pump water into the upper reservoir. When there is higher demand, water is released back into the lower reservoir through a turbine, generating electricity. Reversible turbine/generator assemblies act as a combined pump and turbine generator unit (usually a Francis turbine design).

Types: Natural or Man-made Reservoirs

In open-loop systems, pure pumped-storage plants store water in an upper reservoir with no natural inflows, while pump-back plants utilize a combination of pumped storage and conventional hydroelectric plants with an upper reservoir that is replenished in part by natural inflows from a stream or river. Plants that do not use pumped-storage are referred to as conventional hydroelectric plants; conventional hydroelectric plants that have significant storage capacity may be able to play a similar role in the electrical grid as pumped storage by deferring output until needed.

Economic Efficiency

Taking into account evaporation losses from the exposed water surface and conversion losses, energy recovery of 70-80% or more can be achieved. This technique is currently the most cost-effective means of storing large amounts of electrical energy, but capital costs and the presence of appropriate geography are critical decision factors in selecting pumped-storage plant sites.

The relatively low energy density of pumped storage systems requires either large flows and/or large differences in height between reservoirs. The only way to store a significant amount of energy is by having a large body of water located relatively near, but as high above as possible, a second body of water. In some places this occurs naturally, in others one or both bodies of water were man-made. Projects in which both reservoirs are artificial and in which no natural inflows are involved with either reservoir are referred to as "closed loop" systems.

These systems may be economical because they flatten out load variations on the power grid, permitting thermal power stations such as coal-fired plants and nuclear power plants that provide base-load electricity to continue operating at peak efficiency, while reducing the need for "peaking" power plants that use the same fuels as many base-load thermal plants, gas and oil, but have been designed for flexibility rather than maximal efficiency. Hence pumped storage systems are crucial when coordinating large groups of heterogeneous generators. Capital costs for pumped-storage plants are relatively high, although this is somewhat mitigated by their long service life of up to 75 years or more, which is three to five times longer than utility-scale batteries.

The upper reservoir (Llyn Stwlan) and dam of the festiniog Pumped Storage Scheme in north Wales. The lower power station has four water turbines which generate 360 MW of electricity within 60 seconds of the need arising.

Along with energy management, pumped storage systems help control electrical network frequency and provide reserve generation. Thermal plants are much less able to respond to sudden changes in electrical demand, potentially causing frequency and voltage instability. Pumped storage plants, like other hydroelectric plants, can respond to load changes within seconds.

The most important use for pumped storage has traditionally been to balance baseload powerplants, but may also be used to abate the fluctuating output of intermittent energy sources. Pumped storage provides a load at times of high electricity output and low electricity demand, enabling additional system peak capacity. In certain jurisdictions, electricity prices may be close to zero or occasionally negative on occasions that there is more electrical generation available than there is load available to absorb it; although at present this is rarely due to wind or solar power alone, increased wind and solar generation will increase the likelihood of such occurrences. It is particularly likely that pumped storage will become especially important as a balance for very large scale photovoltaic generation. Increased long distance transmission capacity combined with significant amounts of energy storage will be a crucial part of regulating any large-scale deployment of intermittent renewable power sources. The high non-firm renewable electricity penetration in some regions supplies 40% of annual output, but 60% may be reached before additional storage is necessary.

Small-scale Facilities

While smaller scale pumped storage experiences an economy of scale penalty, there are small-scale installations of such technology, including a recent 13 MW project in Germany. Shell Energy has proposed a 5 MW project in the U.S. state of Washington. Some have proposed small pumped storage plants in buildings, although these are economically unfeasible given the economies of scale present. Also, a large volume of water is required for a meaningful storage capacity which is a difficult fit for an urban setting. Nevertheless, some authors defend its technological simplicity and secure provision of water as important externalities.

Worldwide Use

Adam Beck Complex.

In 2009, world pumped storage generating capacity was 104 GW, while other sources claim 127 GW, which comprises the vast majority of all types of utility grade electric storage. The EU had 38.3 GW net capacity (36.8% of world capacity) out of a total of 140 GW of hydropower and representing 5% of total net electrical capacity in the EU. Japan had 25.5 GW net capacity (24.5% of world capacity).

In 2010 the United States had 21.5 GW of pumped storage generating capacity (20.6% of world capacity). PSH generated (net) -5.501 GWh of energy in 2010 in the US because more energy is consumed in pumping than is generated. Nameplate pumped storage capacity had grown to 21.6 GW by 2014, with pumped storage comprising 97% of grid-scale energy storage in the US. As of late 2014, there were 51 active project proposals with a total of 39 GW of new nameplate capacity across all stages of the FERC licensing process for new pumped storage hydroelectric plants in the US, but no new plants were currently under construction in the US at the time.

The five largest operational pumped-storage plants are listed below:

Station	Country	Location	Capacity (MW)
Bath County Pumped Storage Station	United States	38°12′32″N 79°48′00″W38.20889°N 79.80000°W	3,003
Guangdong Pumped Storage Power Station	China	23°45′52″N 113°57′12″E23.76444°N 113.95333°E	2,400

Huizhou Pumped Storage Power Station	China	23°16′07″N 114°18′50″E23.26861°N 114.31389°E	2,400
Okutataragi Pumped Storage Power Station	Japan	35°14′13″N 134°49′55″E35.23694°N 134.83194°E	1,932
Ludington Pumped Storage Power Plant	United States	43°53′37″N 86°26′43″W43.89361°N 86.44528°W	1,872

Countries with the largest power pumped-storage hydro capacity in 2017	
Country	Power capacity (GW)
China	32
Japan	28.3
United States	22.6
Spain	8
Italy	7.1
India	6.8
Switzerland	6.4
France	5.8

In June 2018 the Australian federal government announced that 14 sites had been identified in Tasmania for pumped storage hydro, with the potential of adding 4.8GW to the national grid if a second interconnector beneath Bass Strait was constructed.

Pump-back Hydroelectric Dams

Conventional hydroelectric dams may also make use of pumped storage in a hybrid system that both generates power from water naturally flowing into the reservoir as well as storing water pumped back to the reservoir from below the dam. The Grand Coulee Dam in the US was expanded with a pump-back system in 1973. Existing dams may be repowered with reversing turbines thereby extending the length of time the plant can operate at capacity. Optionally a pump back powerhouse such as the Russell Dam (1992) may be added to a dam for increased generating capacity. Making use of an existing dam's upper reservoir and transmission system can expedite projects and reduce costs.

In January 2019, the State Grid Corporation of China announced plans to invest US$5.7 billion in five pumped hydro storage plants with a total 6GW capacity, to be located in Hebei, Jilin, Zhejiang, Shandong provinces, and in Xinjiang Autonomous Region. China is seeking to build 40GW of pumped hydro capacity installed by 2020.

Potential Technologies

Seawater

Pumped storage plants can operate with seawater, although there are additional challenges compared to using fresh water. Inaugurated in 1966, the 240 MW Rance tidal power station in France can partially work as a pumped-storage station. When high tides occur at off-peak hours, the turbines can be used to pump more seawater into the reservoir than the high tide would have naturally brought in. It is the only large-scale power plant of its kind.

In 1999, the 30 MW Yanbaru project in Okinawa was the first demonstration of seawater pumped storage. It has since been decommissioned. A 300 MW seawater-based Lanai Pumped Storage Project was considered for Lanai, Hawaii, and seawater-based projects have been proposed in Ireland. A pair of proposed projects in the Atacama Desert in northern Chile would use 600 MW of photovoltaic solar (Skies of Tarapacá) together with 300 MW of pumped storage (Mirror of Tarapacá) raising seawater 600 metres (2,000 ft) up a coastal cliff.

Underground Reservoirs

The use of underground reservoirs has been investigated. Recent examples include the proposed Summit project in Norton, Ohio, the proposed Maysville project in Kentucky (underground limestone mine), and the Mount Hope project in New Jersey, which was to have used a former iron mine as the lower reservoir. The proposed energy storage at the Callio site in Pyhäjärvi (Finland) would utilize the deepest base metal mine in Europe, with 1,450 metres (4,760 ft) elevation difference. Several new underground pumped storage projects have been proposed. Cost-per-kilowatt estimates for these projects can be lower than for surface projects if they use existing underground mine space. There are limited opportunities involving suitable underground space, but the number of underground pumped storage opportunities may increase if abandoned coal mines prove suitable.

In Bendigo, Victoria, Australia, the Bendigo Sustainability Group has proposed the use of the old gold mines under Bendigo for Pumped Hydro Energy Storage. Bendigo has the greatest concentration of deep shaft hard rock mines anywhere in the world with over 5,000 shafts sunk under Bendigo in the second half of the 19th Century. The deepest shaft extends 1,406 metres vertically underground. A recent pre-feasibility study has shown the concept to be viable with a generation capacity of 30 MW and a run time of 6 hours using a water head of over 750 metres.

Decentralised Systems

Small pumped-storage hydropower plants can be built on streams and within infrastructures, such as drinking water networks and artificial snow making infrastructures.

Such plants provide distributed energy storage and distributed flexible electricity production and can contribute to the decentralized integration of intermittent renewable energy technologies, such as wind power and solar power. Reservoirs that can be used for small pumped-storage hydropower plants could include natural or artificial lakes, reservoirs within other structures such as irrigation, or unused portions of mines or underground military installations. In Switzerland one study suggested that the total installed capacity of small pumped-storage hydropower plants in 2011 could be increased by 3 to 9 times by providing adequate policy instruments.

Underwater Reservoirs

In March 2017 the research project StEnSea (Storing Energy at Sea) announced their successful completion of a four-week test of a pumped storage underwater reservoir. In this configuration a hollow sphere submerged and anchored at great depth acts as the lower reservoir, while the upper reservoir is the enclosing body of water. Electricity is created when water is let in via a reversible turbine integrated into the sphere. During off-peak hours the turbine changes direction and pumps the water out again, using "surplus" electricity from the grid. The quantity of power created when water is let in grows proportionally to the height of the column of water above the sphere, in other words: the deeper the sphere is located, the more potential energy it can store, which can be transformed into electric power. On the other hand, pumping the water back out at greater depths also uses up more power, since the turbine-turned-pump must act on the same entire column of water.

As such the energy storage capacity of the submerged reservoir is not governed by the gravitational energy in the traditional sense, but rather by the vertical pressure variation.

While StEnSea's test took place at a depth of 100 m in the fresh water Lake Constance, the technology is foreseen to be used in salt water at greater depths. Since the submerged reservoir needs only a connecting electrical cable, the depth at which it can be employed is limited only by the depth at which the turbine can function, currently limited to 700 m. The challenge of designing salt water pumped storage in this underwater configuration brings a range of advantages:

- No land area is required.
- No mechanical structure other than the electrical cable needs to span the distance of the potential energy difference.
- In the presence of sufficient seabed area multiple reservoirs can scale the storage capacity without limits.
- Should a reservoir collapse, the consequences would be limited apart from the loss of the reservoir itself.

- Evaporation from the upper reservoir has no effect on the energy conversion efficiency.
- Transmission of electricity between the reservoir and the grid can be established from a nearby offshore wind farm limiting transmission loss and obviating the need for onshore cabling permits.

A current commercial design featuring a sphere with an inner diameter of 30 m submerged to 700 m would correspond to a 20 MWh capacity which with a 5 MW turbine would lead to a 4-hour discharge time. An energy park with multiple such reservoirs would bring the storage cost to around a few eurocents per kWh with construction and equipment costs in the range €1,200-€1,400 per kW. To avoid excessive transmission cost and loss, the reservoirs should be placed off deep water coasts of densely populated areas, such as Norway, Spain, USA and Japan. With this limitation the concept would allow for worldwide electricity storage of close to 900 GWh.

For comparison, a traditional, gravity-based pumped storage capable of storing 20 MWh in a water reservoir the size of a 30 m sphere would need a hydraulic head of 519 m with the elevation spanned by a pressurized water pipe requiring typically a hill or mountain for support.

Home Use

Pico Hydro.

Using a pumped-storage system of cisterns and small generators, pico hydro may also be effective for "closed loop" home energy generation systems.

RUN-OF-THE-RIVER HYDROELECTRICITY

Run-of-river hydroelectricity (ROR) or run-of-the-river hydroelectricity is a type of hydroelectric generation plant whereby little or no water storage is provided. Run-of-the-river power plants may have no water storage at all or a limited amount of storage, in which case the storage reservoir is referred to as pondage. A plant without pondage is subject to seasonal river flows, thus the plant will operate as an intermittent energy source. Conventional hydro uses reservoirs, which regulate water for flood control and dispatchable electrical power.

Chief Joseph Dam near Bridgeport, Washington, USA, is a major run-of-the-river station without a sizeable reservoir.

A small and floating run-of-the-river power plant in Austria.

Run-of-the-river or ROR hydroelectricity is considered ideal for streams or rivers that can sustain a minimum flow or those regulated by a lake or reservoir upstream. A small dam is usually built to create a headpond ensuring that there is enough water entering the penstock pipes that lead to the turbines which are at a lower elevation. Projects with pondage, as opposed to those without pondage, can store water for daily load demands. In general, projects divert some or most of a river's flow (up to 95% of mean annual discharge) through a pipe and/or tunnel leading to electricity-generating turbines, then return the water back to the river downstream.

ROR projects are dramatically different in design and appearance from conventional hydroelectric projects. Traditional hydro dams store enormous quantities of water in reservoirs, sometimes flooding large tracts of land. In contrast, run-of-river projects do not have the disadvantages associated with reservoirs, which is why they have less environmental impact.

Saint Marys Falls - run of the river (1902).

The use of the term "run-of-the-river" for power projects varies around the world. Some may consider a project ROR if power is produced with no water storage while limited storage is considered ROR by others. Developers may mislabel a project ROR to soothe public perception about its environmental or social effects. The Bureau of Indian Standards describes run-of-the-river hydroelectricity as:

> "A power station utilizing the run of the river flows for generation of power with sufficient pondage for supplying water for meeting diurnal or weekly fluctuations of demand. In such stations, the normal course of the river is not materially altered."

Many of the larger ROR projects have been designed to a scale and generating capacity rivaling some traditional hydro dams. For example, the Beauharnois Hydroelectric Generating Station in Quebec is rated at 1,853 MW. Some run of the river projects are downstream of other dams and reservoirs. The run of the river project didn't build the reservoir, but does take advantage of the water supplied by it. An example would be the 1995 1,436 MW La Grande-1 generating station. Previous upstream dams and reservoirs are part of the 1980s James Bay Project.

Advantages

When developed with care to footprint size and location, ROR hydro projects can create sustainable energy minimizing impacts to the surrounding environment and nearby communities. Advantages include:

Cleaner Power, Fewer Greenhouse Gases

Like all hydro-electric power, run-of-the-river hydro harnesses the natural potential energy of water, eliminating the need to burn coal or natural gas to generate the electricity needed by consumers and industry. Moreover, run-of-the-river hydro-electric plants do not have reservoirs thus eliminating the methane and carbon dioxide emissions caused by the decomposition of organic matter in the reservoir of a conventional hydro-electric dam. This is a particular advantage in tropical countries where methane generation can be a problem.

Less Flooding/Reservoirs

Without a reservoir, flooding of the upper part of the river does not take place. As a result, people remain living at or near the river and existing habitats are not flooded. Any pre-existing pattern of flooding will continue unaltered, presenting a flood risk to the facility and downstream areas.

Disadvantages

Unfirm Power

Run-of-the-River power is considered an "unfirm" source of power: a run-of-the-river project has little or no capacity for energy storage and hence can't co-ordinate the output of electricity generation to match consumer demand. It thus generates much more power during times when seasonal river flows are high (i.e., spring freshet), and depending on location, much less during drier summer months or frozen winter months.

Availability of Sites

Rapids can provide enough hydraulic head.

The potential power at a site is a result of the head and flow of water. By damming a river, the head is available to generate power at the face of the dam. Where a dam may create a reservoir hundreds of kilometres long, in run of the river the head is usually delivered by a canal, pipe or tunnel constructed upstream of the power house. Due to the cost of upstream construction, a steep drop is desirable, such as falls or rapids.

Environmental Impacts

Small, well-sited ROR projects can be developed with minimal environmental impacts. Larger projects have more environmental concerns. In the case of fish-bearing rivers a ladder may be required and dissolved gases downstream may affect fish.

In British Columbia the mountainous terrain and wealth of big rivers have made it a global testing ground for 10–50 MW run-of-river technology. As of March 2010, there were 628 applications pending for new water licences solely for the purposes of power generation – representing more than 750 potential points of river diversion.

Concerns

- Diverting large amounts of river water reduces river flows, affecting water velocity and depth, reducing habitat quality for fish and aquatic organisms; reduced flows can lead to excessively warm water for salmon and other fish in summer.
- In undeveloped areas, new access roads and transmission lines can cause habitat fragmentation, allowing the introduction of invasive species.
- The lack of reservoir storage may result in intermittent operation, reducing the project's viability.

Major Examples

- Belo Monte Dam, 11,233 megawatts (15,064,000 hp), Pará, Brazil.
- Chief Joseph Dam, 2,620 megawatts (3,510,000 hp).
- Beauharnois Hydroelectric Power Station, Quebec, Canada, 1,903 megawatts (2,552,000 hp).
- Bonneville Dam, 1,092 megawatts (1,464,000 hp).
- Satluj Jal Vidyut Nigam Ltd, Satluj River, Shimla, India, 1,500 megawatts (2,000,000 hp).
- Ghazi-Barostha Hydropower Project on River Indus in Pakistan, 1,450 megawatts (1,940,000 hp).
- La Grande-1 generating station, Quebec, Canada, 1,436 megawatts (1,926,000 hp).
- Kohala Hydropower Project, Jhelum River, Muzaffarabad, Pakistan, 1,100 megawatts (1,500,000 hp).
- Neelum–Jhelum Hydropower Plant, Jhelum River, Muzaffarabad, Azad Kashmir, Pakistan, 969 megawatts (1,299,000 hp).
- Baglihar Hydroelectric Power Projection the Chenab River in India, 900 megawatts (1,200,000 hp).
- Carillon Generating Station, Quebec, Canada, 752 megawatts (1,008,000 hp).
- Upper Tamakoshi Project, Nepal, 456 MW.

- Nyagak Hydroelectric Power Station on Nyagak River in Zombo District, Uganda, 3.5 megawatts (4,700 hp).
- East Toba/Montrose Hydro Project, British Columbia, Canada, 196 megawatts (263,000 hp).
- Forrest Kerr Hydro Project, British Columbia, Canada, 195 megawatts (261,000 hp).
- Patrind Hydropower Plant, Kunhar River, Pakistan, 150 megawatts (200,000 hp).
- Upper Toba Valley, British Columbia, Canada, 123 megawatts (165,000 hp).
- Upper Kotmale Hydropower Project, Talawakele, Sri Lanka, 150 megawatts (200,000 hp).
- Kukule Ganga Power Station, Kelinkanda, Sri Lanka, 75 megawatts (101,000 hp).
- Sechelt Creek Generating Station, British Columbia, Canada, 16 megawatts (21,000 hp).

References

- "AID Foundation 2007 Ashden Award". Archived from the original on 2008-05-28. Retrieved 2008-07-09
- Hydroelectricity, entry: newworldencyclopedia.org, Retrieved 5 February, 2019
- Imeche (Institution of Mechanical Engineers),Hydropower Developments: New Projects, Rehabilitation, and Power Recovery, John Wiley & Sons, 2005 ISBN 1860584799
- Waterwheel-design, hydro-energy: alternative-energy-tutorials.com, Retrieved 3 May, 2019
- Historic England. "Water Street hydraulic power station (1254724)". National Heritage List for England. Retrieved 30 May 2011
- Isenberg, Andrew (2005). Mining California An Ecological History. Hill and Wang. P. 34. ISBN 978-0-8090-9535-3

3

Hydroelectricity Generation

Generation of hydroelectricity generally takes place in hydroelectric power plants, which convert the kinetic energy of water into electricity. Some of the major components of hydroelectric power plants are dam, turbine and generator. This chapter has been carefully written to provide an easy understanding of these facets of hydroelectricity generation.

DAM

Dam is a structure built across a stream, a river, or an estuary to retain water. Dams are built to provide water for human consumption, for irrigating arid and semiarid lands, or for use in industrial processes. They are used to increase the amount of water available for generating hydroelectric power, to reduce peak discharge of floodwater created by large storms or heavy snowmelt, or to increase the depth of water in a river in order to improve navigation and allow barges and ships to travel more easily. Dams can also provide a lake for recreational activities such as swimming, boating, and fishing. Many dams are built for more than one purpose; for example, water in a single reservoir can be used for fishing, to generate hydroelectric power, and to support an irrigation system. Water-control structures of this type are often designated multipurpose dams.

Auxiliary works that can help a dam function properly include spillways, movable gates, and valves that control the release of surplus water downstream from the dam. Dams can also include intake structures that deliver water to a power station or to canals, tunnels, or pipelines designed to convey the water stored by the dam to far-distant places. Other auxiliary works are systems for evacuating or flushing out silt that accumulates in the reservoir, locks for permitting the passage of ships through or around the dam site, and fish ladders (graduated steps) and other devices to assist fish seeking to swim past or around a dam.

A dam can be a central structure in a multipurpose scheme designed to conserve water resources on a regional basis. Multipurpose dams can hold special importance in developing countries, where a single dam may bring significant benefits related to hydroelectric power production, agricultural development, and industrial growth. However, dams have become a focus of environmental concern because of their impact on migrating fish and riparian ecosystems. In addition, large reservoirs can inundate vast tracts of land that are home to many people, and this has fostered opposition to dam projects by groups who question whether the benefits of proposed projects are worth the costs.

In terms of engineering, dams fall into several distinct classes defined by structural type and by building material. The decision as to which type of dam to build largely depends on the foundation conditions in the valley, the construction materials available, the accessibility of the site to transportation networks, and the experiences of the engineers, financiers, and promoters responsible for the project. In modern dam engineering, the choice of materials is usually between concrete, earthfill, and rockfill. Although in the past a number of dams were built of jointed masonry, this practice is now largely obsolete and has been supplanted by concrete. Concrete is used to build massive gravity dams, thin arch dams, and buttress dams. The development of roller-compacted concrete allowed high-quality concrete to be placed with the type of equipment originally developed to move, distribute, and consolidate earthfill. Earthfill and rockfill dams are usually grouped together as embankment dams because they constitute huge mounds of earth and rock that are assembled into imposing man-made embankments.

Rise of Environmental and Economic Concerns

The effect of dams on the natural environment became an issue of public concern at the end of the 20th century. Much of this concern was energized by fears that dams were destroying the populations of migrating (or spawning) fish, which were being blocked or impeded by the construction of dams across rivers and waterways. In more general terms, dams were often perceived—or portrayed—as not simply transforming the environment to serve human desires but also obliterating the environment and causing the destruction of flora and fauna and picturesque landscapes on a massive scale. Dams were also blamed for inundating the cultural homelands of native peoples, who were forced to relocate out of reservoir “take” areas created by large-scale dams. None of these concerns sprang up without warning, and they all have roots that date back many decades.

The environmental problems associated with dams have been exacerbated as dams have increased in height. However, even relatively small dams have prompted opposition by people who believe that their interests are adversely affected by a particular structure. For example, in colonial America, legal action was often taken by upstream landowners who believed that the pond impounded by a small mill dam erected downstream was flooding—and thus rendering unusable—land that could otherwise be used for growing crops or as pasture for livestock. By the late 18th century, when many mill dams were beginning to reach heights that could not easily be jumped or traversed by spawning fish, some people sought to have them removed because of their effect on fishing. In such situations, opposition to dams is not driven by an abstract concern for the environment or the survival of riparian ecosystems; rather, it is driven by an appreciation that a particular dam is transforming the environment in ways that serve only certain special interests.

In the 1870s one of the first wide-scale efforts to block the construction of a dam because of misgivings about its potential effect upon the landscape came in the Lake District of northwestern England. The Lake District is recognized as one of the most picturesque regions of England because of its mountains and rolling hills. However, this same landscape also offered a good location for an artificial reservoir that could feed high-quality water to the growing industrial city of Manchester almost 160 km (100 miles) to the south. The city's Thirlmere Dam was eventually built and generally accepted as a positive development, but not before it aroused impassioned opposition among citizens throughout the country who feared that part of England's natural and cultural heritage might be defiled by the creation of a "water tank" in the midst of the Lake District.

In the United States a similar but even more impassioned battle erupted in the early 20th century over plans by the city of San Francisco to build a reservoir in Hetch Hetchy Valley. Located more than 900 metres (3,000 feet) above sea level, the Hetch Hetchy site offered a good storage location in the Sierra Nevada for water that could be delivered without pumping to San Francisco via an aqueduct nearly 270 km (167 miles) long. Hetch Hetchy, however, is also located within the northern boundaries of Yosemite National Park. The renowned naturalist John Muir led the way in fighting the proposed dam and—with assistance from Sierra Club members and other citizens across the United States who were concerned about the loss of natural landscapes to commercial and municipal development—made the fight over the preservation of Hetch Hetchy Valley a national issue. In the end, the benefits to be provided by the dam—including the development of at least 200,000 kilowatts of hydroelectric power—outweighed the costs to be exacted by the inundation of the valley. Approved by the U.S. Congress in 1913, the construction of the dam, known today as O'Shaughnessy Dam in honour of the city engineer who oversaw its construction, was a defeat for the Sierra Club and landscape preservationists, who continued to use it as a symbol and rallying cry for mid-20th-century environmental causes.

After World War II, plans were made by the U.S. Bureau of Reclamation to build a hydroelectric power dam across the Green River at Echo Park Canyon within the boundaries of Dinosaur National Monument in eastern Utah. Many of the same issues raised at Hetch Hetchy were again debated, but in this instance opponents such as the Sierra Club were able to block construction of the dam through a concerted effort to lobby Congress and win support from the American public at large. However, in its effort to save Echo Park, the Sierra Club dropped opposition to the proposed Glen Canyon Dam across the Colorado River near the Arizona-Utah border, and this 216-metre (710-foot) high concrete arch dam, built between 1956 and 1966, eventually came to be seen by environmentalists as being responsible for destroying a beautiful pristine landscape encompassing thousands of square kilometres. Anger over the Glen Canyon Dam energized the Sierra Club to mount a major campaign against additional dams proposed for construction along the Colorado River near the borders of Grand Canyon National Park. By the late 1960s, plans for these proposed Grand Canyon dams were politically dead. Although the reasons for their demise were largely the result of regional water conflicts between states in the Pacific Northwest and states in the American Southwest, the environmental movement took credit for saving America from the desecration of a national treasure.

Glen Canyon DamConstruction of the Glen Canyon Dam on the Colorado River formed Lake Powell in Arizona.

In developing parts of the world, dams are still perceived as an important source of hydroelectric power and irrigation water. Environmental costs associated with dams have nonetheless attracted attention. In India the relocation of hundreds of thousands of people out of reservoir areas generated intense political opposition to some dam projects.

Xiling Gorge, in the Three Gorges section of the Yangtze River (Chang Jiang), as it appeared before completion of the Three Gorges Dam, Hubei province, China.

In China the Three Gorges Dam (constructed from 1994 to 2006) generated significant opposition within China and in the international community. Millions of people were displaced by, and cultural and natural treasures were lost beneath, the reservoir that was created following erection of the 185-metre- (607-foot-) high concrete wall, some 2,300 metres (7,500 feet) long, across the Yangtze River. The dam is capable of producing 22,500 megawatts of electricity (which can reduce coal usage by millions of tons per year), making it one of the largest hydroelectric producers in the world.

Dams still unquestionably have an important role to play within the world's social, political, and economic framework. But for the foreseeable future, the specific character of that role and the way that dams will interrelate with the environment will likely remain a subject of contentious debate.

The Modern Dam

Basic Problems in Dam Design

Most modern dams are of two basic types: masonry (concrete) gravity designs and embankment (earthfill or rockfill) designs. Masonry dams are typically used to block streams running through relatively narrow gorges, as in mountainous terrain; although the structures may be very high, the total amount of material required for such sites is limited. Embankment dams are often preferred to control rivers and streams passing through broad, wide valleys where only a very long barrier, requiring a great volume of material, will suffice. The choice of design depends on the geology and configuration of the site, the purposes of the dam, and cost factors related to material supply and site accessibility.

Site Investigation and Testing

Investigation of a site for a dam includes sinking trial borings to determine geological strata. These borings can be supplemented by shafts and tunnels. In the shafts and tunnels, which are often used sparingly because of their cost, tests can be made to measure strength, elasticity, permeability, and prevailing stresses in rock strata, with particular attention given to the properties of thin partings, or walls, between the more massive beds. The presence in groundwater of chemical solutions harmful to the materials to be used in the construction of the dam must be assessed. Sources of construction materials (such as sand and rock aggregate needed in the production of concrete) often require exploration. As a design increases in height, the study of foundation conditions becomes more important because the pressures that will be exerted on the foundation increase proportionally.

Model tests can play a major role in the structural, seismic, and hydraulic design of dams. Structural models can be particularly useful in analysis of arch dams and in verifying analytical stress calculations. Various materials have been used for model tests; for example, rubber was used on some early tests for Hoover Dam. The need for accurate reproduction of stress patterns in complex models is met by using material of low elasticity. In a sense, dams themselves are models for future design, and large-scale test dams were built as far back as the 1920s. The instruments built into them to record movements under load, strains (or deformations) that occur within various parts of the dam under reservoir loadings, temperature and pressure changes, and other factors are installed primarily to study the performance of the structure and to warn of possible emergencies, but their value in confirming design assumptions is important.

Computers have permitted considerable advances in computational and analytic methods of design. Their ability to handle great volumes of data and to solve large sets of simultaneous equations containing many variables made the finite-element method practicable. In this method a complicated structure is divided into a number of separate equilibrium conditions, and strains (or deflections) are rendered compatible, thus leading to a complete analysis of stress and strain distribution throughout the structure. However, computers only model or approximate conditions as they exist in the real world and are not a substitute for judicious engineering judgment during the design process.

Problems of Materials

Each of the two basic dam materials, concrete and earthfill, possesses weaknesses that must be accommodated in the design process.

Weaknesses of Concrete

Unless reinforced with embedded steel bars, concrete is weak in tensile strength; that is, it can easily crack or be pulled apart. Concrete dams are therefore designed to place minimum tensile stress on the dam and instead to take advantage of concrete's great compressive strength. The chief constituent of concrete, cement, shrinks as it hardens, and it also releases heat as part of the chemical reactions that occur within the cement during the process of hydration (or hardening). Because of the massive quantities of concrete used in a large dam, shrinkage caused by cooling can present a serious cracking hazard.

Various expedients are used to counter the likelihood of cracking, and much attention is often paid to reducing the amount of heat generated by the concrete. Concrete is usually cast (or poured) in separate, distinct blocks with heights (or "lifts") of no more than about 1.5 metres (5 feet). Gaps between these blocks may be left to facilitate heat dispersal, and these gaps can be filled in later with cement grout. Low-heat cements may also be used, and these are specially blended so that the production of heat by the setting concrete is minimized. In the interior portions of a massive concrete dam, where impermeability or strength in resisting climatic and chemical deterioration are not particularly important attributes, the amount of cement in the concrete mix can be reduced; in turn, this reduces the heat generated. The cement content, and therefore the heat caused by hydrating, can also be reduced by using aggregate consisting of large stones. It is also possible to use fine-grained materials, such as fly ash (pulverized fuel), as filler, reducing the total cement volume in the concrete. Another technique is to use air-entraining agents that permit using a lower water-to-cement ratio in mixing the concrete. Techniques used to speed the cooling process include replacing some of the water in the mix by ice, circulating cool water through pipes placed within the concrete (this technology was used to great advantage during the construction of Hoover Dam), and extracting excess water from surfaces by vacuuming.

Weaknesses of Earthfill

Compared with concrete, soils and rock fragments lack strength, are much more permeable, and possess less resistance to deterioration and disturbance by flowing water. These disadvantages are compensated for by a much lower cost and by the ability of earthfill to adapt to deformation caused by movements in the dam foundation. This assumes, of course, sufficient usable soil or rockfill is available near the dam site. Earthfill is often quite economical, provided that a suitable "borrow" area can be utilized close to the construction site.

Soil consists of solid particles with water and air in between. When the soil is compressed by loading, as occurs in dam construction, some drainage of air and water takes place, causing an increase in pressures between the solid particles. When there is a high rate of seepage, the soil tends to develop differential pressures and reach a condition called quick, in which it behaves as a fluid. Even if it does not reach this condition, there is often some weakening of its structure, and steps must be taken to counter this.

The Earthquake Problem

Many large dams have been built in the seismically active regions of the world, including Japan, the western United States, New Zealand, the Himalayas, and the Middle East. In 1968 the Tokachi earthquake damaged 93 dams in Honshu, the main Japanese island; all were embankment dams of relatively small height.

Despite a great deal of work on the distribution of seismic activity, the measurement of strong ground motions, and the response of dams to such motions, earthquake design of dams remains imprecise. The characteristics of strong ground motions at a given site cannot be predicted, and all types of dams possess some degree of freedom, imperfect elasticity, and imprecise damping. Nevertheless, computers and model testing offer the promise of future continued progress. It is now possible to calculate the response of a concrete dam to any specified ground motion; this has been done for the Tang-e Soleyman Dam in Iran and the Gariep Dam in South Africa.

Because the foundations of concrete dams are typically keyed into bedrock, concrete dams usually do not experience great accelerations when shaken by earthquakes; for this reason, concrete dams have achieved an excellent safety record in terms of withstanding seismic forces. The safety record for embankment dams is also good, with the notable exception of earthfill dams constructed using hydraulic fill technology. Such dams retain a large quantity of water within their soil structure, which renders them vulnerable to liquefaction of the saturated soil when hit by a seismic shock. In 1971 the Van Norman Dam (or lower San Fernando Dam) in Los Angeles suffered partial collapse when a large quantity of hydraulic fill "slipped" during an earthquake. In recent years, engineers have also come to appreciate that large artificial reservoirs can trigger earthquakes that would not occur in the absence of the reservoirs. Reservoir-induced earthquakes may be caused by the extra weight of the water or, more typically, by increases in the groundwater pore pressure reducing the strength of the rock beneath the reservoir. These tremors are usually not large, but they can cause minor damage to communities in the region surrounding the dam.

Types of Dams

The Modern Concrete Dam

Concrete Gravity Dams

Concrete gravity dams usually run in a straight line across a broad valley and resist the horizontal thrust of the retained water entirely by their own weight. The three main forces acting on a gravity dam are the thrust of the water stored in the reservoir, the weight of the dam, and the pressure exerted by the foundation. It is also essential to consider the thrust exerted on the upstream face by silt deposited in the reservoir or by ice on the water surface, the inertial forces that can be caused by seismic action, and, in particular, the buoyant uplift force of water seeping under the dam or into the horizontal joints.

Uplift from seepage has caused sustained discussion among engineers dating back as far as the 1890s. Uplift calls for the greatest of care in design and construction. Where a dam is founded on solid rock, a simple downward projection of concrete into the rock will generally suffice to cut off seepage and eliminate uplift pressures. Usually, however, the rock foundation is permeable, sometimes to considerable depths, so construction of an absolutely reliable cutoff is either difficult or impossible. Reliance must then be placed on an extensive system of grouting the fissured rock and on relieving uplift pressures by means of drainage. Many dams possess both cutoffs and underdrainage.

Another development in the construction of gravity dams is incorporation of posttensioned steel into the structure. For example, this helped reduce the cross section of Allt na Lairige Dam in Scotland to only 60 percent of that of a conventional gravity dam of the same height. A series of vertical steel rods near the upstream water face, stressed by jacks and securely anchored into the rock foundation, resists the overturning tendency of this more slender section. This system has also been used to raise existing gravity dams to a higher crest level, economically increasing the storage capacity of a reservoir.

Of special interest are three concrete gravity dams that feature a straight sloping downstream face. Bratsk, built across the Angara River at Irkutsk in Russia and completed in 1964, stands 125 metres (410 feet) above foundation level and, excluding the earthen side dams, is nearly 1,525 metres (5,000 feet) in length; it contains 4,500,000 cubic metres (5,900,000 cubic yards) of concrete. Grand Coulee Dam, completed in 1941, was built across the Columbia River in Washington state, U.S.; its main structure is 168 metres (550 feet) high and 1,592 metres (5,223 feet) long and contains almost 9,000,000 cubic metres (12,000,000 cubic yards) of concrete. Grande Dixence Dam in Switzerland, completed in 1962 across the narrower valley of the Dixence, has a crest length of 700 metres (2,296 feet) and contains approximately 5,960,000 cubic metres (7,790,000 cubic yards) of concrete; at 285 metres (935 feet) it was the highest dam in the world until the Nurek Dam on the Vakhsh River in Tajikistan was completed in 1980, with a height of 317 metres (1,040 feet). By comparison, the Great Pyramid of Giza in Egypt contains 2,600,000 cubic metres (3,400,000 cubic yards) of masonry.

Concrete Buttress and Multiple-arch Dams

Unlike gravity dams, buttress dams do not rely entirely upon their own weight to resist the thrust of the water. Their upstream face, therefore, is not vertical but inclines about 25° to 45°, so the thrust of the water on the upstream face inclines toward the foundation. Embryonic buttresses existed in some Roman dams built in Spain, among them the Proserpina. As technology advanced, dams with thin buttresses of reinforced concrete supporting an inclined upstream face were built. In today's buttress dams, less account is taken of effecting maximum economy in the use of concrete. The trend is to reduce the area of costly formwork necessary and to avoid use of steel reinforcement. With greater heights, modern buttress dams are inevitably less slender.

Several variations are possible in the design of the junction between the buttresses at the water face. Where no relative movement in the buttress foundations is anticipated, the design can link individual buttress heads rigidly, by means of arches, to form a multiple-arch dam. A Canadian example of this type is the 214-metre- (703-foot-) high multiple-arch Daniel Johnson Dam on the Manicouagan River in Quebec. The dam, which was completed in 1968, uses a total of 14 buttresses in its crest length of 1,310 metres (4,297 feet); two very much larger buttresses support the structure over the original riverbed.

Where buttress foundations might yield, the design must allow some freedom of movement between the heads of the buttresses. This is normally achieved by enlarging the heads until they are almost in contact and then joining them with flexible seals. Thus joined, the heads present a solid face to the water. Such a design was used in the construction of the Farahnaz Pahlavi Dam in Iran. Built for the Tehrān Regional Water Board in 1967, this dam has a maximum height of 107 metres (351 feet) and a crest length of nearly 360 metres (1,181 feet).

A comparison between the Daniel Johnson multiple-arch dam and the Farahnaz Pahlavi buttress dam shows that the buttresses have to be placed much closer together than is necessary with a multiple-arch dam. This allows each buttress to be more slender, however, and spreads the load more evenly over the foundation. The detailed design at the bottom of the Farahnaz Pahlavi buttresses was necessitated by weak foundation conditions at the site and by the need to limit the length of each buttress to reduce its response to seismic action. By contrast, the Daniel Johnson buttresses could be founded individually, exploiting fully an important advantage of buttress dams over gravity dams—that of smaller uplift forces.

Arch Dams

The advantages of building a curved dam—thus using the water pressure to keep the joints in the masonry closed—were appreciated as early as Roman times. An arch dam is a structure curving upstream, where the water thrust is transferred either directly to the valley sides or indirectly through concrete abutments. Theoretically, the ideal constant angle arch in a V-shaped valley has a central angle of 133° of curvature. This led to the development of the "constant-angle" (or variable radius) arch dam, first built at Salmon Creek in Alaska in 1913–14.

An arch dam is a thick shell structure that derives strength from its curved profile. Dependent for its strength upon effective support at its abutments, its very strength and rigidity make it sensitive to movements at the abutments. Only favourable sites providing sound rock are suitable for arch dams.

The great reserves of strength inherent in an arch dam were dramatically displayed in 1963 when the reservoir behind Vaiont Dam in Italy was virtually destroyed by a landslide. Vaiont, at that time the second highest dam in the world, was built across a narrow gorge on limestone foundations so that the crest, 262 metres (858 feet) above the valley bottom, was only 190 metres (623 feet) in length. Some large-scale instability in the mountainside above the reservoir had been observed earlier by the engineers during filling; they were allowed to proceed very slowly, and three years later, on Oct. 9, 1963, with filling still incomplete, about 240 million cubic metres (314 million cubic yards) of soil and rock slid down into the reservoir, sending a tremendous volume of water to a height of 260 metres (853 feet) on the opposite side of the valley. The flood overtopped the dam to a depth of 100 metres (328 feet) and surged down the valley, destroying several villages and causing large loss of life. Yet only superficial damage was caused to the dam, which is about 3.4 metres (11.2 feet) thick at its crest.

Embankment Dams

Early embankments of earthfill or rockfill were often built as simple homogeneous structures, with the same material used throughout. No effort was made at first to subdivide the dam into separate zones with the best-suited material in each zone. Like a concrete gravity dam, the weight of an embankment dam deflects the horizontal thrust of the water pressure down to the foundation. The resultant pressures on the foundation must not cause excessive deformation, as this will result in failure.

Unlike concrete, embankment dam materials possess only limited resistance to water penetration. The rate of penetration depends on the pressures exerted by the water in the reservoir, the length of seepage paths through the dam, and the permeability of the material of construction. Soils and rock range from substantially impermeable clays through silts and sands to coarse-graded gravels and rock fragments that possess little resistance to the movement of water. The range is extremely wide; the seepage rate through clean gravel is 10,000 times that through sand, 10,000,000 times that through silt, and 100,000,000 times that through dense clay.

An embankment dam must be stable, and its side slopes must not slip or slide. In addition, liquefaction of the soils must not occur, and erosion of the soils—as a result of water overtopping the crest, wave action on the upstream face, or seepage washing out the finer material through the coarser—must be avoided. As with a concrete dam, seepage of water from the reservoir through the foundation and under the actual embankment also must be controlled in order to ensure safety.

Potential Weakness

There are three parts of a dam where weakening of the soil structure and liquefactions can occur. In the top figure the pattern of seepage through a homogeneously filled dam is shown. Near the downstream toe, the gradient of the pore water pressures is steep, and constraints holding the soil structure together are low; this is one area of weakness in an embankment dam. One solution is to introduce drainage as indicated in the bottom figure, where the area of steep seepage gradients has been moved to where the soil is constrained near the centre of the dam.

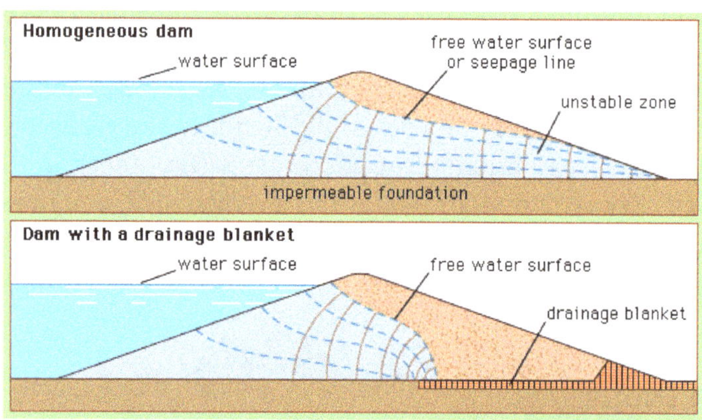

Paths of seepage through embankment dams.

A second area of potential weakness is the upstream face, when the water in the reservoir is rapidly drawn down. If the pore water pressures cannot adjust themselves fast enough to this change in the free water surface in the reservoir, severe seepage gradients begin; these can cause failure. A zone of freely draining fill of coarser grading can be placed on the upstream face to counter this.

Water seepage from the reservoir through the foundations under the dam is another potential weakness. Because of their great base widths, embankment dams can be constructed on unfavourable sites, such as open-joined rock or weaker and possibly locally permeable clay. It is necessary, however, either to check or to harmlessly drain away the seepage water that would otherwise weaken the downstream parts of the dam and, in extreme cases, cause it to fail. Several countermeasures, possibly in combination, can be employed: the foundation can be grouted or a cutoff trench excavated and backfilled with an impermeable material; a drainage blanket can be constructed at the base of the downstream part of the dam, or individual drainage wells or galleries can be excavated; the length of the seepage paths under the dam can be extended by means of an impermeable blanket laid on the upstream side of the dam; or additional free-draining fill can be placed at the downstream toe of the dam.

Construction Techniques

Today all large embankment dams have a core of lower permeability built near their centre. The width of the core is restricted to that necessary to lower sufficiently the pore pressures in the downstream part of the dam. Although the top of the core must be at the crest of the dam, the core itself need not be vertical. On some rockfill dams the core can slope forward to an extreme position where it lies on the upstream face.

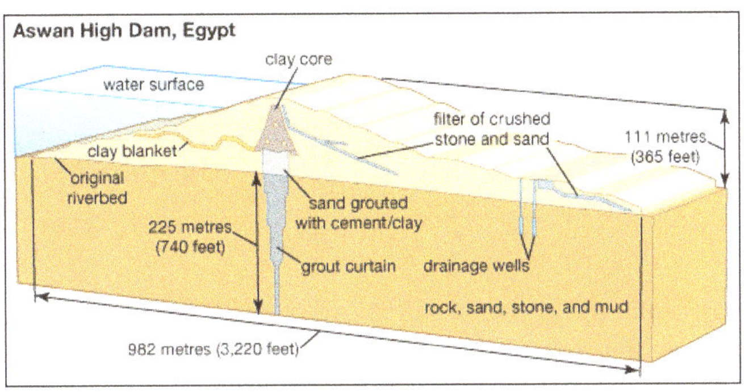

Aswan High Dam, Egypt.

Where seepage is inevitable, the use of finely graded core material in proximity to coarser material is avoided. Bands of intermediately graded material must be inserted to prevent the finely graded material from leaching through the coarse zones. Filter zones are graded so each band is four to five times coarser than the preceding band.

A typical section of the Aswan High Dam in Egypt, which was completed in 1970, shows an embankment 111 metres (365 feet) high built of dune sand and rockfill on a very permeable foundation of deep alluvium. There the central clay core is vertical; this barrier to seepage is extended to the original riverbed as grouted sand and below the riverbed to a depth of 225 metres (740 feet) as a grout curtain. A corrugated blanket of clay extends upstream within the dam from the base of the core. Within the upstream and downstream cofferdams, partly of rockfill, much of the filling is of compacted sand. Filter layers separate the cofferdam filler from the outer layers of freely draining rockfill. Drainage wells are observed below the downstream toe. The early stages of construction were carried out under deep water—hence the use of grouted coarse sand between the clay core and the grout curtain.

Until completion of the 317-metre- (1,040-foot-) high Nurek Dam in Tajikistan in 1980, Oroville Dam in California (completed in 1968) was the highest embankment dam in the world, at 236 metres (774 feet). Unlike the Aswan High Dam, Oroville was not built on deep permeable alluvium, nor was it necessary to place part of the fill underwater. One unusual feature is the concrete block at the base of the sloping core designed to fill in the incised gorge of the Feather River Canyon. The grout curtain, compared with that of the Aswan High Dam, is of nominal depth. On each side of the sloping core, transition zones separate the core from the main mass of more pervious filling. The downstream transition zone is backed by a curtain drain of selected pervious material connected to a drainage blanket on the downstream side. The upstream face of the dam is protected against wave action by a 1-metre (3-foot) layer of broken stone (riprap).

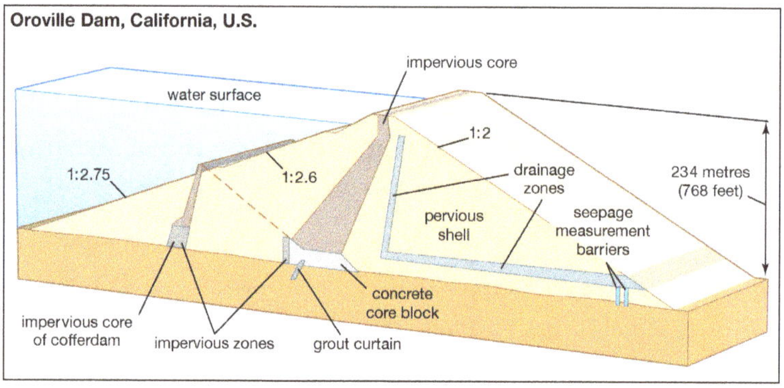

Oroville Dam, California.

Efficient compacting of soils requires maximum density of dry particles consistent with an economic number of passes of the compacting plant. The process of compacting a soil by kneading it involves expelling as much of the air as practicable; water content is not normally much reduced. The optimum water content for maximum dry density—which results in maximum strength—can be achieved for a given amount of work done on the soil in compaction. In arid climates, water must often be added to excavated soils. In temperate climates, however, water content is usually too high, except in deeply excavated and well-drained soils.

Normally, soils are placed in embankment dams in thin layers individually compacted by rolling. Finer soils, such as those used in cores, may be harrowed before rolling. Coarser soils, including rock fragments, are compacted by vibration and then rolled. Coarse rock fragments (rockfill) are compacted to a limited extent by impact on being dumped from the construction plant. In the process of hydraulic filling, sands are dredged from borrow pits, transported in water by pipelines to the filling area, and deposited there by draining off the surplus water. Hydraulic filling is widely practiced in maritime works, and it has also been used for embankment dams. In the early 20th century it was a widely used construction technique for dams, but the practice fell out of favour after the Fort Peck Dam across the Missouri River in northeastern Montana experienced a partial failure during construction in the late 1930s.

Auxiliary Structures

Spillways

Serious consequences can follow if a dam is overtopped. Disaster is likely in the case of an embankment dam not designed to permit uncontrolled flow of water on its downstream slope. In March 1960 the partially completed embankment dam at Orós, Brazil, was accidentally overtopped during a period of unexpectedly heavy rainfall. Despite heroic efforts to avert disaster, the water level rose nearly 1 metre (3 feet) above crest level, eroded about half the fill in the dam, and cut a deep breach about 200 metres (660 feet) wide in the structure. Although there was time to evacuate 100,000 people living downstream, half were rendered homeless and about 50 perished. Spillage over a concrete gravity dam is also serious, because the floodwater erodes the foundations at the downstream toe. Arch dams possess greater resistance to failure after overtopping.

Flood hydrology is a difficult subject to precisely quantify, but much effort is being made to establish relationships between rainfall and river discharge. Although statistical methods cannot determine the maximum possible flood, they can indicate the probability of a specified flow being exceeded in a particular period. For example, engineers found that, in constructing the Kariba Dam over the Zambezi River on the border between Zambia and Zimbabwe, analyses of the available records of river discharge yielded the estimate that a flood of 7,600 cubic metres (9,950 cubic yards) per second should be expected once in four years. During the first year of construction on the riverbed, a flood of 8,500 cubic metres (11,100 cubic yards) per second was experienced, and in the second year the Zambezi discharged 16,200 cubic metres (21,200 cubic yards) per second.

Kariba Dam, on the Zambezi River at the border between Zambia and Zimbabwe.

In these circumstances, civil engineers attach much importance to the design of spillways on dams. Inadequate spillway capacity caused failure by overtopping for many older earthen dams built before modern flood data became available.

Four general aspects of spillways are worth noting. First, the uncontrolled discharge of surplus water past the dam should be automatic and not dependent upon human control. Second, the spillway intake should be wide enough so that the largest floods can pass without increasing the water level in the reservoir enough to cause a nuisance to upstream property owners. Third, the rate of floodwater discharge should not increase much above that experienced before the construction of the dam. An increase in discharge can cause flood problems downstream, but a dam usually reduces the peak discharge rate because of the lag effect caused by a flood passing through the reservoir. Fourth, floodwater discharged over the height of a dam can be destructive to the dam structure itself and to the riverbed unless its energy is controlled and dissipated in harmless turbulence.

With embankment dams, a separate spillway structure is normally constructed to one side of the dam. With concrete gravity dams, the sloping downstream face of the structure can often serve as the basis for the spillway. Water flowing down a spillway can travel at very high speeds—about 160 km (100 miles) per hour in the case of a dam 100 metres (330 feet) high—and form a standing wave where it enters the riverbed; it proceeds downstream at lower mean velocity but in a highly turbulent state. Grand Coulee Dam utilizes a spillway of this type. An obstruction known as a kicker, placed at the toe of the dam to project the water slightly upward, can move farther downstream the area in which erosion of the riverbed is most intense. With higher dams it is possible to deflect the jet of spilling water from a level above the base of the dam; this is known as a ski-jump spillway.

Spillways need not be open to the atmosphere. Shaft and tunnel spillways can carry away the water to a point downstream of the dam. At the upstream end, the intake can be self-priming siphons or bell-mouthed drop shafts; the latter are also known as morning-glory spillways.

With arch dams it is convenient to construct gated openings in the shell structure at some distance below the crest of the dam, ensuring that the discharging jets fall well clear downstream. A line of six such gates is used in the design of Kariba Dam.

Spillways constructed to one side of earthen dams are featured in the design of Oroville Dam and of Mangla Dam in Pakistan. The spillway at Mangla discharges 28,000 cubic metres (36,600 cubic yards) of water per second; the upper stilling basin has the dimensions of an Olympic Games stadium, including its grandstands.

Gates

In addition to spillways, openings through dams are also required for drawing off water for irrigation and water supply, for ensuring a minimum flow in the river for riparian interests downstream, for generating power, and for evacuating water and silt from the reservoir. These gated openings normally are fitted with coarse screens at the upstream ends to prevent entry of floating and submerged debris. Provision for cleaning these screens is essential.

Several forms of gates have been developed. The simplest and oldest form is a vertical-lift gate that, sliding or rolling against guides, can be raised to allow water to flow underneath. Radial, or tainter, gates are similar in principle but are curved in vertical section to better resist water pressure. Tilting gates consist of flaps held by hinges along their lower edges that permit water to flow over the top when they are lowered.

Drum gates can control the reservoir level upstream to precise levels automatically and without the assistance of mechanical power. One drum gate design consists of a shaped-steel caisson held in position by hinges mounted on the crest of the dam and supported in a flotation chamber constructed immediately downstream of the crest. Water pressure in the reservoir and buoyancy of the caisson in the flotation chamber hold the caisson in rotational equilibrium. Raising or lowering the water level in the flotation chamber causes the caisson to rotate in the same direction, thus reducing or increasing flow from the reservoir over the gate. This action can be linked to and operated automatically by a float control device in the reservoir. Two drum gates are installed at Pitlochry Dam in Scotland.

Reservoirs

Modern engineers have learned the value of giving attention early to potential problems in reservoir maintenance. Sediment in rivers seriously influences the effective life of a reservoir and therefore the financing of a dam. Some modern dams have been rendered useless for storing water because the reservoir has filled with silt. In many others, effective storage capacity has been seriously reduced. At the Nile barrages, the heavy silt-laden floodwater is allowed to pass through the sluices so that only the cleaner water at the end of the flood season is stored.

Fish Passes

For centuries people have appreciated that dams can have dramatic effects on fish populations, but concern about this issue increased significantly starting in the 1930s, with the construction of major dams along the Columbia River and its tributaries in the Pacific Northwest. Success in accommodating fish runs has been achieved with salmon in Scotland and on certain rivers in the United States and Canada. Notable examples of conservation measures are to be found at Bonneville Dam, along the lower Columbia River, and at many dams in Scotland.

Bonneville Dam stems the Columbia River in Washington and Oregon.
The dam's fish ladders help salmon swim upstream to their spawning grounds.

Adult salmon swimming to their spawning grounds upstream must be prevented by screens from entering the turbine tailraces at power stations and induced instead to enter a fish pass that allows them to surmount the dam. Similarly, young salmon must be allowed to pass a dam safely on their journey downstream to feeding grounds in the ocean. Young salmon are surprisingly insensitive to sudden changes of pressure and have been known to pass safely through turbines operating at heads of up to 49 metres (160 feet). Nevertheless, it is preferable to induce them to use the fish passes.

Fish passes usually take the form of fish ladders and fish locks. A fish ladder is utilized at Pitlochry Dam in Scotland; it consists of a series of stepped pools through which water is continuously discharged during the migratory seasons. The individual pools may be separated by a series of low weirs or linked by short inclined underwater pipes to provide the necessary steps of less than a metre in water levels. Sometimes both weirs and pipes are provided.

The Borland fish lock was developed in Scotland as an alternative to fish ladders. It operates on the same intermittent principle as a ship lock but is constructed as a closed conduit. Intermittent closure of the gates at the bottom causes the continuous flow through the lock to fill the conduit at intervals, which allows fish waiting in the bottom chamber to be raised through the height of the dam. The lock also serves at other seasons to flush young salmon down past the dam.

Unfortunately, as more dams were built along some rivers, the success of fish ladders and other technologies designed to obviate the effects of dams proved difficult to sustain, and migrating fish began to experience dramatic population declines on several rivers. The most notable of these declines came on the Lower Snake River (a tributary of the Columbia River), where a series of dams built in the 1960s and '70s to generate hydroelectric power and to make the port of Lewiston, Idaho, accessible to ocean-going barges were tagged by environmentalists as causing the destruction (and possible extinction) of many native spawning fish. In the 1990s these dams on the Lower Snake River were the focus of widely publicized dam-removal initiatives. In other parts of the

United States, relatively small dams (most notably the Edwards Dam across the lower Kennebec River in Maine) have been removed from rivers in order to help revitalize spawning fish populations.

HYDROELECTRIC POWER PLANT

Hydropower plants capture the energy of falling water to generate electricity. A turbine converts the kinetic energy of falling water into mechanical energy. Then a generator converts the mechanical energy from the turbine into electrical energy.

Hydroplants range in size from "micro-hydros" that power only a few homes to giant dams like Hoover Dam that provide electricity for millions of people.

Parts of a Hydroelectric Plant

Most conventional hydroelectric plants include four major components:

- Dam: Raises the water level of the river to create falling water. Also controls the flow of water. The reservoir that is formed is, in effect, stored energy.
- Turbine: The force of falling water pushing against the turbine's blades causes the turbine to spin. A water turbine is much like a windmill, except the energy is provided by falling water instead of wind. The turbine converts the kinetic energy of falling water into mechanical energy.
- Generator: Connected to the turbine by shafts and possibly gears so when the turbine spins it causes the generator to spin also. Converts the mechanical energy from the turbine into electric energy. Generators in hydropower plants work just like the generators in other types of power plants.
- Transmission lines: Conduct electricity from the hydropower plant to homes and business.

Electricity Produced by a Hydroelectric Plant

The amount of electricity a hydropower plant produces depends on two factors:

- How Far the Water Falls: The farther the water falls, the more power it has. Generally, the distance that the water falls depends on the size of the dam. The higher the dam, the farther the water falls and the more power it has. Scientists would say that the power of falling water is "directly proportional" to the distance it falls. In other words, water falling twice as far has twice as much energy.
- Amount of Water Falling: More water falling through the turbine will produce more power. The amount of water available depends on the amount of water flowing down the river. Bigger rivers have more flowing water and can produce more energy. Power is also "directly proportional" to river flow. A river with twice the amount of flowing water as another river can produce twice as much energy.

Generation of electricity by hydropower (potential energy in stored water) is one of the cleanest methods of producing electric power. In 2012, hydroelectric power plants contributed about 16% of total electricity generation of the world. Hydroelectricity is the most widely used form of renewable energy. It is a flexible source of electricity and also the cost of electricity generation is relatively low.

Layout and Working of Hydroelectric Power Plant

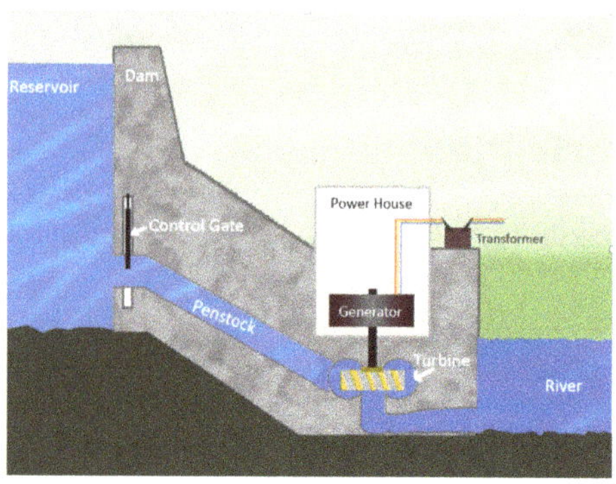

The above image shows the typical layout of a hydroelectric power plant and its basic components.

Dam and Reservoir: The dam is constructed on a large river in hilly areas to ensure sufficient water storage at height. The dam forms a large reservoir behind it. The height of water level (called as water head) in the reservoir determines how much of potential energy is stored in it.

Control Gate: Water from the reservoir is allowed to flow through the penstock to the turbine. The amount of water which is to be released in the penstock can be controlled by a control gate. When the control gate is fully opened, maximum amount of water is released through the penstock.

Penstock: A penstock is a huge steel pipe which carries water from the reservoir to the turbine. Potential energy of the water is converted into kinetic energy as it flows down through the penstock due to gravity.

Water Turbine: Water from the penstock is taken into the water turbine. The turbine is mechanically coupled to an electric generator. Kinetic energy of the water drives the turbine and consequently the generator gets driven. There are two main types of water turbine; (i) Impulse turbine and (ii) Reaction turbine. Impulse turbines are used for large heads and reaction turbines are used for low and medium heads.

Generator: A generator is mounted in the power house and it is mechanically coupled to the turbine shaft. When the turbine blades are rotated, it drives the generator and electricity is generated which is then stepped up with the help of a transformer for the transmission purpose.

Surge Tank

Surge tanks are usually provided in high or medium head power plants when considerably long penstock is required. A surge tank is a small reservoir or tank which is open at the top. It is fitted between the reservoir and the power house. The water level in the surge tank rises or falls to reduce the pressure swings in the penstock. When there is sudden reduction in load on the turbine, the governor closes the gates of the turbine to reduce the water flow. This causes pressure to increase abnormally in the penstock. This is prevented by using a surge tank, in which the water level rises to reduce the pressure. On the other hand, the surge tank provides excess water needed when the gates are suddenly opened to meet the increased load demand.

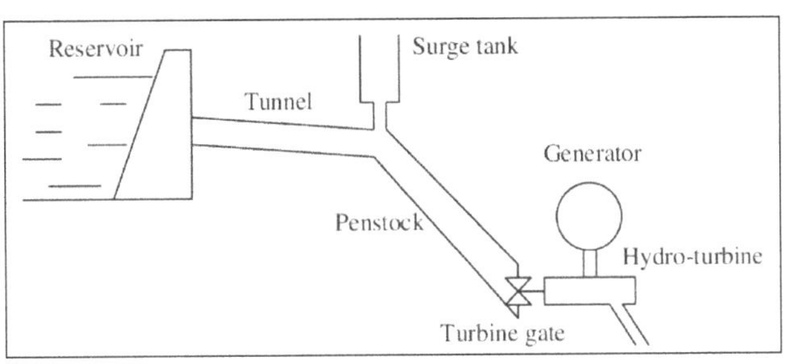

Types of Hydro-Power Plants

Conventional Plants

Conventional plants use potential energy from dammed water. The energy extracted depends on the volume and head of the water. The difference between height of water level in the reservoir and the water outflow level is called as water head.

Pumped Storage Plant

In pumped storage plant, a second reservoir is constructed near the water outflow from the turbine. When the demand of electricity is low, the water from lower reservoir is pumped into the upper (main) reservoir. This is to ensure sufficient amount of water available in the main reservoir to fulfil the peak loads.

Run-of-River Plant

In this type of facility, no dam is constructed and, hence, reservoir is absent. A portion of river is diverted through a penstock or canal to the turbine. Thus, only the water flowing from the river is available for the generation. And due to absence of reservoir, any oversupply of water is passed unused.

Advantages of a Hydroelectric Power Plant

- No fuel is required as potential energy is stored water is used for electricity generation.
- Neat and clean source of energy.
- Very small running charges - as water is available free of cost.
- Comparatively less maintenance is required and has longer life.
- Serves other purposes too, such as irrigation.

Disadvantages

- Very high capital cost due to construction of dam.
- High cost of transmission – as hydro plants are located in hilly areas which are quite away from the consumers.

WATERMILL

A watermill or water mill is a mill that uses hydropower. It is a structure that uses a water wheel or water turbine to drive a mechanical process such as milling (grinding), rolling, or hammering. Such processes are needed in the production of many material goods, including flour, lumber, paper, textiles, and many metal products. These watermills may comprise gristmills, sawmills, paper mills, textile mills, hammermills, trip hammering mills, rolling mills, wire drawing mills.

One major way to classify watermills is by wheel orientation (vertical or horizontal), one powered by a vertical waterwheel through a gear mechanism, and the other equipped with a horizontal waterwheel without such a mechanism. The former type can be further divided, depending on where the water hits the wheel paddles, into undershot, overshot, breastshot and pitchback (backshot or reverse shot) waterwheel mills. Another way to classify water mills is by an essential trait about their location: tide mills use the movement of the tide; ship mills are water mills onboard (and constituting) a ship.

Watermills impact the river dynamics of the watercourses where they are installed. During the time watermills operate channels tend to sedimentate, particularly backwater. Also in the backwater area, inundation events and sedimentation of adjacent floodplains increase. Over time however these effects are cancelled by river banks becoming higher. Where mills have been removed, river incision increases and channels deepen.

Watermill of Braine-le-Château, Belgium.

Interior of the Lyme Regis watermill, UK.

Operation

Typically, water is diverted from a river or impoundment or mill pond to a turbine or water wheel, along a channel or pipe (variously known as a flume, head race, mill race, leat, leet, lade (Scots) or penstock). The force of the water's movement drives the blades of a wheel or turbine, which in turn rotates an axle that drives the mill's other machinery. Water leaving the wheel or turbine is drained through a tail race, but this channel may also be the head race of yet another wheel, turbine or mill. The passage of water is controlled by sluice gates that allow maintenance and some measure of flood control; large mill complexes may have dozens of sluices controlling complicated interconnected races that feed multiple buildings and industrial processes.

Watermills can be divided into two kinds, one with a horizontal water wheel on a vertical axle, and the other with a vertical wheel on a horizontal axle. The oldest of these were horizontal mills in which the force of the water, striking a simple paddle wheel set horizontally in line with the flow turned a runner stone balanced on the rynd which is atop a shaft leading directly up from the wheel. The bedstone does not turn. The problem with this type of mill arose from the lack of gearing; the speed of the water directly set the maximum speed of the runner stone which, in turn, set the rate of milling.

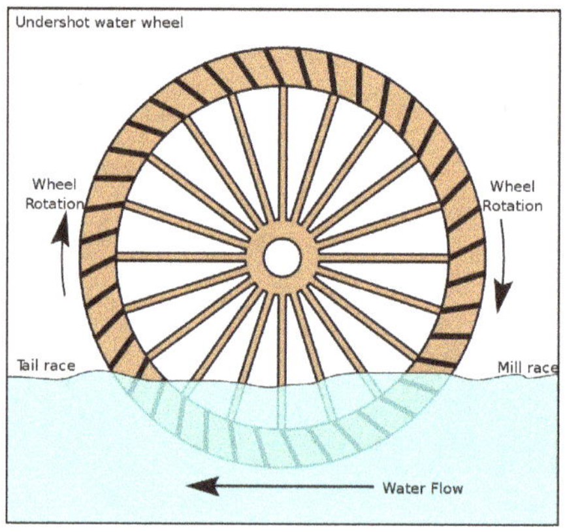

Undershot water wheel.

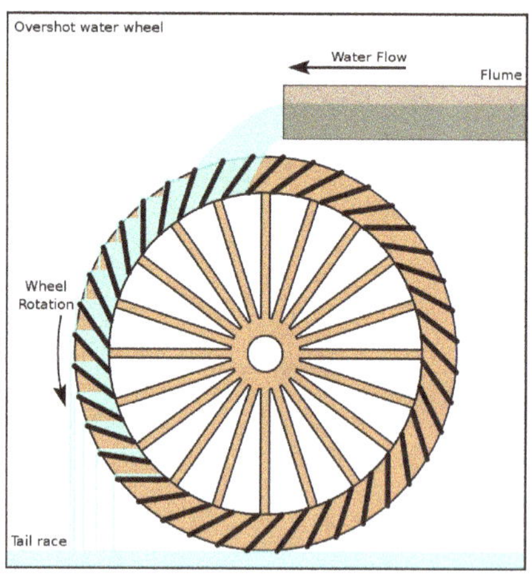

Overshot water wheel.

Most watermills in Britain and the United States of America had a vertical waterwheel, one of four kinds: undershot, breast-shot, overshot and pitchback wheels. This vertical produced rotary motion around a horizontal axis, which could be used (with cams) to lift hammers in a forge, fulling stocks in a fulling mill and so on.

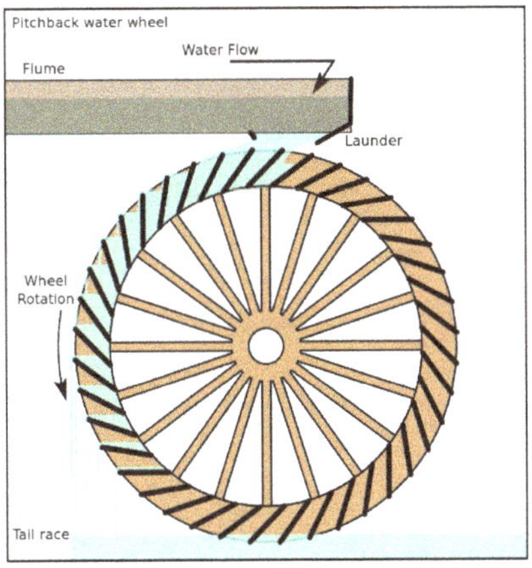

Pitchback water wheel, often used to increase the power generated by a breastshot wheel.

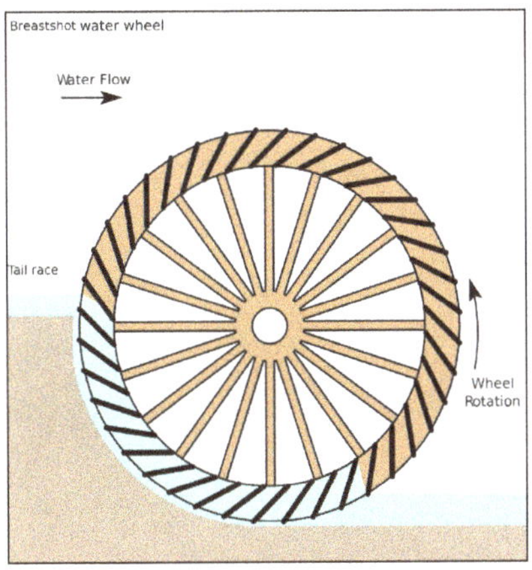

Breastshot water wheel.

Milling Corn

However, in corn mills rotation about a vertical axis was required to drive its stones. The horizontal rotation was converted into the vertical rotation by means of gearing, which also enabled the runner stones to turn faster than the waterwheel. The usual arrangement in British and American corn mills has been for the waterwheel to turn a horizontal shaft on which is also mounted a large pit wheel. This meshes with the wallower, mounted on a vertical shaft, which turns the (larger) great spur wheel (mounted on the same shaft). This large face wheel, set with pegs, in turn, turned a smaller wheel (such as a lantern gear) known as a stone nut, which was attached to the shaft that drove the runner stone. The number of runner stones that could be turned depended directly upon the supply of water available. As waterwheel technology improved mills became more efficient, and by the 19th century, it was common for the great spur wheel to drive several stone nuts, so that a single water wheel could drive as many as four stones. Each step in the process increased the gear ratio which increased the maximum speed of the runner stone. Adjusting the sluice gate and thus the flow of the water past the main wheel allowed the miller to compensate for seasonal variations in the water supply. Finer speed adjustment was made during the milling process by tentering, that is, adjusting the gap between the stones according to the water flow, the type of grain being milled, and the grade of flour required.

Mulino Meraviglia in San Vittore Olona, Italy, along Olona river.

In many mills (including the earliest) the great spur wheel turned only one stone, but there might be several mills under one roof. The earliest illustration of a single water-wheel driving more than one set of stones was drawn by Henry Beighton in 1723 and published in 1744 by J. T. Desaguliers.

Dalgarven Mill, Ayrshire, United Kingdom.

Shipmill on the Mura, Slovenia.

Overshot and Pitchback Mills

The overshot wheel was a later innovation in waterwheels and was around two and a half times more efficient than the undershot. The undershot wheel, in which the main water wheel is simply set into the flow of the mill race, suffers from an inherent inefficiency stemming from the fact that the wheel itself, entering the water behind the main thrust of the flow driving the wheel, followed by the lift of the wheel out of the water ahead of the main thrust, actually impedes its own operation. The overshot wheel solves this problem by bringing the water flow to the top of the wheel. The water fills buckets built into the wheel, rather than the simple paddle wheel design of undershot wheels. As the buckets fill, the weight of the water starts to turn the wheel. The water spills out of the bucket on the down side into a spillway leading back to river. Since the wheel itself is set above the spillway, the water never impedes the speed of the wheel. The impulse of the water on the wheel is also harnessed in addition to the weight of the water once in the buckets. Overshot wheels require the construction of a dam on the river above the mill and a more elaborate millpond, sluice gate, mill race and spillway or tailrace.

An inherent problem in the overshot mill is that it reverses the rotation of the wheel. If a miller wishes to convert a breastshot mill to an overshot wheel all the machinery in the mill has to be rebuilt to take account of the change in rotation. An alternative solution was the pitchback or backshot wheel. A launder was placed at the end of the flume on the headrace, this turned the direction of the water without much loss of energy, and the direction of rotation was maintained. Daniels Mill near Bewdley, Worcestershire is an example of a flour mill that originally used a breastshot wheel, but was converted to use a pitchback wheel. Today it operates as a breastshot mill.

A breastshot waterwheel at Dalgarven Mill, United Kingdom.

Larger water wheels (usually overshot steel wheels) transmit the power from a toothed annular ring that is mounted near the outer edge of the wheel. This drives the machinery using a spur gear mounted on a shaft rather than taking power from the central axle. However, the basic mode of operation remains the same; gravity drives machinery through the motion of flowing water.

Toward the end of the 19th century, the invention of the Pelton wheel encouraged some mill owners to replace over- and undershot wheels with Pelton wheel turbines driven through penstocks.

Tide Mills

A different type of watermill is the tide mill. This mill might be of any kind, undershot, overshot or horizontal but it does not employ a river for its power source. Instead a mole or causeway is built across the mouth of a small bay. At low tide, gates in the mole are opened allowing the bay to fill with the incoming tide. At high tide the gates are closed, trapping the water inside. At a certain point a sluice gate in the mole can be opened allowing the draining water to drive a mill wheel or wheels. This is particularly effective in places where the tidal differential is very great, such as the Bay of Fundy in Canada where the tides can rise fifty feet, or the now derelict village of Tide Mills, East Sussex. The last two examples in the United Kingdom which are restored to working conditions can be visited at Eling, Hampshire and at Woodbridge, Suffolk.

Run of the river schemes do not divert water at all and usually involve undershot wheels the mills are mostly on the banks of sizeable rivers or fast flowing streams. Other watermills were set beneath large bridges where the flow of water between the stanchions was faster. At one point London bridge had so many water wheels beneath it that bargemen complained that passage through the bridge was impaired.

Current Status

Watermill in Jahodná (Slovakia).

By the early 20th century, availability of cheap electrical energy made the watermill obsolete in developed countries although some smaller rural mills continued to operate commercially later throughout the century. A few historic mills such as the Water Mill, Newlin Mill and Yates Mill in the US and The Darley Mill Centre in the UK still operate for demonstration purposes. Small-scale commercial production is carried out in the UK at Daniels Mill, Little Salkeld Mill and Redbournbury Mill.

Some old mills are being upgraded with modern hydropower technology, such as those worked on by the South Somerset Hydropower Group in the UK.

In some developing countries, watermills are still widely used for processing grain. For example, there are thought to be 25,000 operating in Nepal, and 200,000 in India. Many of these are still of the traditional style, but some have been upgraded by replacing wooden parts with better-designed metal ones to improve the efficiency. For example, the Centre for Rural Technology in Nepal upgraded 2,400 mills between 2003 and 2007.

Applications

Watermill in Caldas Novas, Brazil.

Former watermill in Kohila, Estonia.

- Bark mills ground bark, from oak or chestnut trees to produce a coarse powder for use in tanneries.
- Blade mills were used for sharpening newly made blades.
- Blast furnaces, finery forges, and tinplate works were, until the introduction of

the steam engine, almost invariably water powered. Furnaces and Forges were sometimes called iron mills.

- Bobbin mills made wooden bobbins for the cotton and other textile industries.
- Carpet mills for making carpets and rugs were sometimes water-powered.
- Cotton mills were driven by water. The power was used to card the raw cotton, and then to drive the spinning mules and ring frames. Steam engines were initially used to increase the water flow to the wheel, then as the industrial revolution progressed, to directly drive the shafts.
- Fulling or *walk mills were used for a finishing process on woollen cloth.*
- Gristmills, or *corn mills, grind* grains into flour.
- Lead was usually smelted in smeltmills prior to the introduction of the cupola (a reverberatory furnace).
- Needle mills for scouring needles during manufacture were mostly water-powered (such as Forge Mill Needle Museum).
- Oil mills for crushing oil seeds might be wind or water-powered.
- Paper mills used water not only for motive power, but also required it in large quantities in the manufacturing process.
- Powder mills for making gunpowder - black powder or smokeless powder were usually water-powered.
- Rolling mills shaped metal by passing it between rollers.
- Sawmills cut timber into lumber.
- Slitting mills were used for slitting bars of iron into rods, which were then made into nails.
- Spoke mills turned lumber into spokes for carriage wheels.
- Stamp mills for crushing ore, usually from non-ferrous mines.
- Textile mills for spinning yarn or weaving cloth were sometimes water-powered.

TROMPE

A trompe is a water-powered air compressor, commonly used before the advent of the electric-powered compressor. A trompe is somewhat like an airlift pump working in reverse.

A Catalan forge trompe.

Trompes were used to provide compressed air for bloomery furnaces in Catalonia and the USA. The presence of a trompe is a signature attribute of a Catalan forge, a type of bloomery furnace.

Trompes can be enormous. At Canadian Hydro Developers' Ragged Chute facility in New Liskeard, Ontario, water falls down a shaft 351 feet (107 m) deep and 9 ft (2.7 m) across to generate compressed air for mining equipment and ventilation.

Operation

Trompes are very simple devices. They consist of four main parts: water-supply pipe or shaft with an air-inlet inside it, water outflow pipe, separation chamber and takeoff air-pipe. The vertical pipe or shaft goes down from higher point to a separation chamber; a pipe, that is typically narrower than previous one, coming away from that chamber, allows the water to exit at a lower level, and another pipe (air-pipe) coming from the chamber allows the compressed air to exit as needed.

Water rushing down the vertical pipe falls through a constriction. The constriction produces a lower pressure because of the venturi effect, and an external port allows air to be sucked in thus creating a constant air supply. The air forms bubbles in the pipe. As the bubbles go down the pipe they are pressurized proportionally to the hydraulic head, which is the height of the column of water in the pipe. The compressed air rises to the top of the separation chamber(wind box). The separation chamber has a compressed-air takeoff pipe, and the compressed air can be used as a power source.

The energy of the falling water creates negative pressure inside the pipe that is compensated by the air from the outside atmosphere provided through inlet. The air is compressed by surrounding water pressure (which increases under a column due to the discharge to atmospheric pressure). The pressure of the air delivered cannot exceed the hydraulic head of the discharge pipe of the separation chamber.

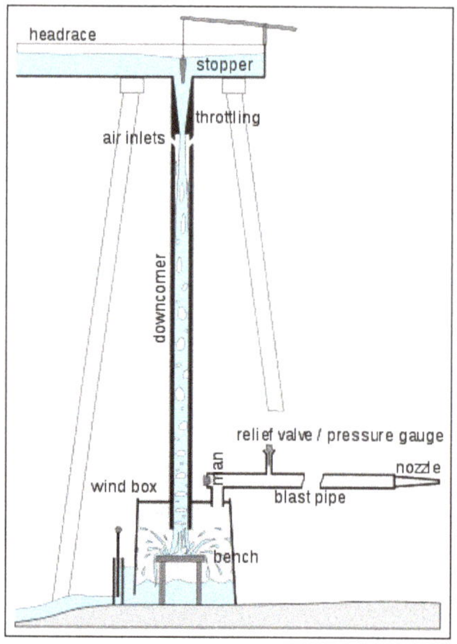

Sketch of a hydraulic trompe with the name of its different constituent elements.

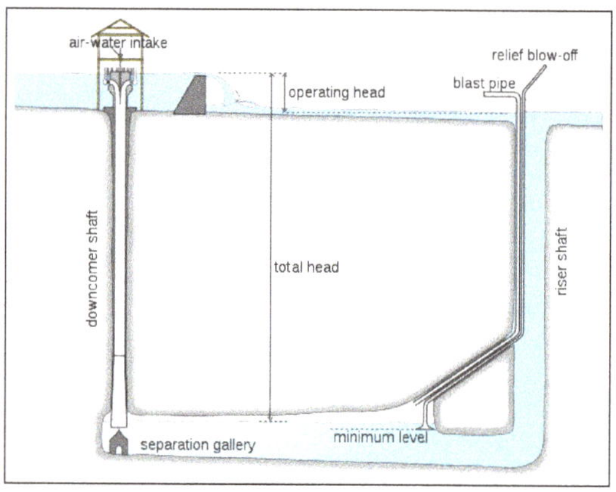

Principle of a Taylor hydraulic trompe.

Large trompes were often situated at high waterfalls so that ample head was available. (However, trompes can raise the water, via siphon-effect, nearly to 70% of its initial elevation.) The Ragged Chute plant on the Montreal River near the town of Cobalt, Ontario, is a trompe and tourist attraction. It is now owned by Canadian Hydro and exists beside a modern hydroelectric plant.

Compressed air from a trompe is at the temperature of the water, and its partial pressure of water vapor is that of the dewpoint of the water's temperature. If the water is cool, the compressed air can be made very dry by passing it through pipes that are warmer than the water. Often, ordinary outside air can warm the pipes enough to produce dry, cool compressed air.

Today, trompes constructed of plastic pipe are being used to provide aeration for mine drainage treatment. In this application, mine water is used to drive the trompe and the compressed air that is generated is used to oxygenate the mine water and to drive off excess dissolved carbon dioxide that may be present thus raising the pH of the water being treated.

The trompe is closely related to the Sprengel vacuum pump which uses mercury falling through a tube to create a vacuum instead of pressure.

HYDRAULIC POWER NETWORK

A hydraulic power network is a system of interconnected pipes carrying pressurized liquid used to transmit mechanical power from a power source, like a pump, to hydraulic equipment like lifts or motors. The system is analogous to an electrical grid transmitting power from a generating station to end-users. Only a few hydraulic power transmission networks are still in use; modern hydraulic equipment has a pump built into the machine. In the late 19th century, a hydraulic network might have been used in a factory, with a central steam engine or water turbine driving a pump and a system of high-pressure pipes transmitting power to various machines.

The idea of a public hydraulic power network was suggested by Joseph Bramah in a patent obtained in 1812. William Armstrong began installing systems in England from the 1840s, using low-pressure water, but a breakthrough occurred in 1850 with the introduction of the hydraulic accumulator, which allowed much higher pressures to be used. The first public network, supplying many companies, was constructed in Kingston upon Hull, England. The Hull Hydraulic Power Company began operation in 1877, with Edward B. Ellington as its engineer. Ellington was involved in most of the British networks, and some further afield. Public networks were constructed in Britain at London, Liverpool, Birmingham, Manchester and Glasgow. There were similar networks in Antwerp, Melbourne, Sydney, Buenos Aires and Geneva. All of the public networks had ceased to operate by the mid-1970s, but Bristol Harbour still has an operational system, with an accumulator situated outside the main pumphouse, enabling its operation to be easily visualised.

Joseph Bramah, an inventor and locksmith living in London, registered a patent at the London Patent Office on 29 April 1812, which was principally about a provision of a public water supply network, but included a secondary concept for the provision of a high-pressure water main, which would enable workshops to operate machinery. The high-pressure water would be applied "to a variety of other useful purposes, to which the same has never before been so applied." Major components of the system were a ring main, into which a number of pumping stations would pump the water, with pressure being regulated by several air vessels or loaded pistons. Pressure relief valves would protect the system, which he believed could deliver water at a pressure of "a great plurality of atmospheres", and in concept, this was how later hydraulic power systems worked.

The pumping station and hydraulic accumulator at Bristol Docks.

In Newcastle upon Tyne, a solicitor called William Armstrong, who had been experimenting with water-powered machines, was working for a firm of solicitors who were appointed to act on behalf of the Whittle Dene Water Company. The water company had been set up to supply Newcastle with drinking water, and Armstrong was appointed secretary at the first meeting of shareholders. Soon afterwards, he wrote to Newcastle Town Council, suggesting that the cranes on the quay should be converted to hydraulic power. He was required to carry out the work at his own expense, but would be rewarded if the conversion was a success. It was, and he set up the Newcastle Cranage Company, which received an order for the conversion of the other four cranes. Further work followed, with the engineer from Liverpool Docks visiting Newcastle and being impressed by a demonstration of the crane's versatility, given by the crane driver John Thorburn, known locally as "Hydraulic Jack".

While the Newcastle system ran on water from the public water supply, the crane installed by Armstrong at Burntisland was not located where such an option was possible, and so he built a 180-foot (55 m) tower, with a water tank at the top, which was filled by a 6 hp (4.5 kW) steam engine. At Elswick in Glasgow, charges by the Corporation Water Department for the water used persuaded the owners that the use of a steam-powered crane would be cheaper. Bramah's concept of "loaded pistons" was introduced in 1850, when the first hydraulic accumulator was installed as part of a scheme for cranes for the Manchester, Sheffield and Lincolnshire Railway. A scheme for cranes at Paddington the following year specified an accumulator with a 10-inch (250 mm) piston and a stroke of 15 feet (4.6 m), which enabled pressures of 600 pounds per square inch (41 bar) to be achieved. Compared to the 80 psi (5.5 bar) of the Newcastle scheme, this increased pressure significantly reduced the volumes of water used. Cranes were not the only application, and hydraulic operation of the dock gates at Swansea reduced the operating time from 15 to two minutes, and the number of men required to operate them from twelve to four. Each of these schemes was for a single customer, and the application of hydraulic power more generally required a new model.

Public Power in the United Kingdom

Machell Street hydraulic pumping station in Hull, showing the water settling tank on the roof.

Kingston Upon Hull

The first practical installation which supplied hydraulic power to the public was in Kingston upon Hull, in England. The Hull Hydraulic Power Company began operation in 1876. They had 2.5 miles (4.0 km) of pipes, which were up to 6 inches (150 mm) in diameter, and ran along the west bank of the River Hull from Sculcoates bridge to its junction with the Humber. The pumping station was near the north end of the pipeline, on Machell Street, near the disused Scott Street bascule bridge, which was powered hydraulically. There was an accumulator at Machell Street, and another one much nearer the Humber, on the corner of Grimsby Lane. Special provision was made where the pressure main passed under the entrance to Queens Dock. By 1895, pumps rated at 250 hp (190 kW) pumped some 500,000 imperial gallons (2,300 m^3) of water into the system each week, and 58 machines were connected to it. The working pressure was 700 psi (48 bar), and the water was used to operate cranes, dock gates, and a variety of other machinery connected with ships and shipbuilding. The Hull system lasted until the 1940s, when the systematic bombing of the city during the Second World War led to the destruction of much of the infrastructure, and the company was wound up in 1947, when Mr F J Haswell, who had been the manager and engineer since 1904, retired.

The man responsible for the Hull system was Edward B. Ellington, who had risen to become the managing director of the Hydraulic Engineering Company, based in Chester, since first joining it in 1869. At the time of its installation, such a scheme seemed like "a leap in the dark", according to R. H. Tweddell writing in 1895, but despite a lack of enthusiasm for the scheme, Ellington pushed ahead and used it as a test bed for both the mechanical and the commercial aspects of the idea. He was eventually involved on some level in most of the hydraulic power networks of Britain. The success of such systems led to them being installed in places as far away as Antwerp in Belgium, Melbourne and Sydney in Australia, and Buenos Aires in Argentina.

Independent hydraulic power networks were also installed at Hull's docks - both the Albert Dock (1869), and Alexandra Dock (1885) installed hydraulic generating stations and accumulators.

London

The best-known public hydraulic network was the citywide network of the London Hydraulic Power Company. This was formed in 1882, as the General Hydraulic Power Company, with Ellington as the consulting engineer. By 1883, another enterprise, the Wharves and Warehouses Steam Power and Hydraulic Pressure Company, had begun to operate, with 7 miles (11 km) of pressure mains on both sides of the River Thames. These supplied cranes, dock gates, and other heavy machinery. Under the terms of an Act of Parliament obtained in 1884, the two companies amalgamated to become the London Hydraulic Power Company. Initially supplying 17.75 million gallons (80.7 megalitres) of high-pressure water each day, this had risen to 1,650 million gallons (7,500 megalitres) by 1927, when the company was powering around 8,000 machines from the supply. They maintained 184 miles (296 km) of mains at 700 psi (48 bar), which covered an area reaching Pentonville in the north, Limehouse in the east, Nine Elms and Bermondsey in the south and Earls Court and Notting Hill in the west.

The preserved pumping equipment in Wapping pumping station, which was owned by the London Hydraulic Power Company.

Five pumping stations kept the mains pressurised, assisted by accumulators. The original station was at Falcon Wharf, Bankside, but this was replaced by four stations at Wapping, Rotherhithe, Grosvenor Road in Pimlico and City Road in Clerkenwell. A fifth station at East India Docks was originally operated by the Port of London Authority, but was taken over and connected to the system. The stations used steam engines until 1953, when Grosvenor Road station was converted to use electric motors, and following the success of this project, the other four were also converted. The electric motors allowed much smaller accumulators to be used, since they were then only controlling the pressure and flow, rather than storing power. While the network supplied lifts, cranes and dockgates, it also powered the cabaret platform at the Savoy Hotel, and from 1937, the 720-tonne three-section central floor at the Earls Court Exhibition Centre, which could be raised or lowered relative to the main floor to convert between a swimming pool and an exhibition hall. The London system contracted during the Second World War, due to the destruction of customers' machinery and premises. Following the hostilities, large areas of London were reconstructed, and the re-routing of pressure mains was much more difficult than the provision of an electric supply, so that by 1954 the number of machines had fallen to 4286. The company was wound up in 1977.

Liverpool

A system began operating in Liverpool in 1888. It was an offshoot of the London-based General Hydraulic Power Company, and was authorised by acts of Parliament obtained in 1884 and 1887. By 1890, some 16 miles (26 km) of mains had been installed, supplied by a pumping station at Athol Street, on the bank of the Leeds and Liverpool Canal. Although water was originally taken from the canal, cleaner water supplied by Liverpool Corporation was in use by 1890, removing the need for a filtration plant. At this time two pumpsets were in use, and a third was being installed. Pressure was maintained by two accumulators, each with an 18-inch (460 mm) diameter piston with a stroke of 20 feet (6.1 m). *The Practical Engineer* quoted the pressure as 75 pounds per square inch (5.2 bar), but this is unlikely to be correct by comparison with other systems. A second pumping station at Grafton Street was operational by 1909. The system ceased operation in 1971.

Birmingham

Birmingham obtained its system in 1891, when the Dalton Street hydraulic station opened. In an unusual move, J. W. Gray, the Water Department engineer for the city, had been laying pressure mains beneath the streets for some years, anticipating the need for such a system. The hydraulic station used Otto 'Silent' type gas engines, and had two accumulators, with an 18-inch (460 mm) diameter piston, a stroke of 20 feet (6.1 m) and each loaded with a 93-tonne weight. The gas engines were started by a small hydraulic engine, which used the hydraulic energy stored in the accumulators, and all equipment was supplied by Ellington's company. Very few documents describing the details of the system are known to exist.

Manchester and Glasgow

The final two public systems in Britain were in Manchester, commissioned in 1894, and Glasgow, commissioned the following year. Both were equipped by Ellington's company, and used the higher pressure of 1,120 psi (77 bar). This was maintained by six sets of triple-expansion steam engines, rated at 200 hp (150 kW) each. Two accumulators with pistons of 18-inch (460 mm) diameter, a stroke of 23 feet (7.0 m), and loaded with 127 tonnes were installed. In Manchester, the hydraulic station was built on the east side of Gloucester Street, by Manchester Oxford Road railway station. It was later supplemented by stations at Water Street and Pott Street, the latter now under the car parks of the Central Retail Park. At its peak in the 1930s, the system consisted on 35 miles (56 km) of pipes, which were connected to 2,400 machines, most of which were used for baling cotton. The system was shut down in 1972. In Glasgow, the pumping station was at the junction of High Street and Rottenrow. By 1899, it was supplying power to 348 machines, and another 39 were in the process of being completed. The pipes were 7 inches (180 mm) in diameter, and there were around 30 miles (48 km) of them by 1909, when 202,141 imperial gallons (918.95 m^3) of high pressure water were supplied to customers. The system was shut down in 1964.

Systems Outside the United Kingdom

Antwerp

All of the British systems were designed to provide power for intermittent processes, such as the operation of dock gates or cranes. The system installed at Antwerp was somewhat different, in that its primary purpose was the production of electricity for lighting. It was commissioned in 1894, and used pumping engines producing a total of 1,000 hp (750 kW) to supply water at 750 psi (52 bar). Ellington, writing in 1895, stated that he found it difficult to see that this was an economical use of hydraulic power, although tests conducted at his works at Chester in October 1894 showed that efficiencies of 59 per cent could be achieved using a Pelton wheel directly coupled to a dynamo.

Australia

Two major systems were built in Australia. The first was in Melbourne, where the Melbourne Hydraulic Power Company began operating in July 1889. The company was authorised by an Act of the Victorian Parliament passed in December 1887, and construction of the system began, with Coates & Co. acting as consulting engineers, and George Swinburne working as engineering manager. The steam pumping plant was supplied by Abbot & Co. from England. Expansion was rapid, with around 70 machines, mainly hydraulic lifts, connected to the system by the end of 1889, and a third steam engine had to be installed in mid-1890, which more than doubled the capacity of the system. A fourth pumping engine was added in 1891, by which time there were 100 customers connected to the mains. The mains were a mixture of 4-inch (100 mm) and 6-inch

(150 mm) pipes. The water was extracted from the Yarra River until 1893, after which it was drawn from the Public Works Department's supply. There were some 16 miles (26 km) of mains by 1897. A second pumping station was added in 1901, and in 1902, 102 million gallons (454 megalitres) of pressurised water were used by customers.

The system was operated as a commercial enterprise until 1925, after which the business and its assets reverted to the City of Melbourne, as specified by the original act. One of the early improvements made by the City Council was to consolidate the system. The steam pumps were replaced by new electric pumps, located in the Spencer Street power station, which thus supplied both electric power and hydraulic power to the city. The hydraulic system continued to operate under municipal ownership until December 1967.

In January 1891, a system in Sydney came on-line, having been authorised by act of Parliament in 1888. George Swinburne was again the engineer, and the system was supplying power to around 200 machines by 1894, which included 149 lifts and 20 dock cranes. The operating company was the Sydney and Suburbs Hydraulic Power Company, later shortened to the Sydney Hydraulic Power Company. Pressure mains were either of 4-inch (100 mm) or 6-inch (150 mm) diameter, and at its peak, there were around 50 miles (80 km) of mains, covering an area between Pyrmont, Woolloomooloo and Broadway. In 1919, most of the 2369 lifts in the metropolitan area were hydraulically operated. The pumping station, together with two accumulators, was situated in the Darling Harbour district, and the original steam engines were replaced by three electric motors driving centrifugal pumps in 1952. The scheme remained in private ownership until its demise in 1975, and the pumping station has since been re-used as a tavern.

Buenos Aires

Ellington's system in Buenos Aires was designed to operate a sewage pumping scheme in the city.

Geneva

Geneva created a public system in 1879, using a 300 hp (220 kW) steam engine installed at the Pont de la Machine to pump water from Lake Geneva, which provided drinking water and a pressurized water supply for the city. The water power was used by about a hundred small workshops having Schmid type water engines installed. The power of the engines was between 1 and 4 hp (0.75 and 2.98 kW) and the water was supplied at a pressure of 2 to 3 bars (29 to 44 psi).

Due to increased demand a new pumping plant was installed which started operation in 1886. The pumps were driven by Jonval turbines using the water power of the river Rhône. This structure was called Usine des Forces Motrices and was one of the largest structures for generation and distribution of power at the time of construction. By 1897, a total of 18 turbines had been installed, with a combined rating of 3.3MW.

The distribution network used three different pressure levels. The drinking water supply used the lowest pressure, while the intermediate and the high pressure mains served as hydraulic power networks. The intermediate pressure mains operated at 6.5 bars (94 psi) and by 1896 some 51 miles (82 km) of pipework had been installed. It was used for powering 130 Schmid type water engines with a gross power of 230 hp (170 kW). The high pressure network had an operating pressure of 14 bars (200 psi) bar and had a total length of 58 miles (93 km). It was used to power 207 turbines and motors, as well as elevator drives, and had a gross power of 3,000 hp (2,200 kW).

Many turbines were used for driving generators for electric lighting. In 1887 an electricity generation plant was built next to the powerhouse, which generated 110 V DC with a maximum power of 800 hp (600 kW) and an AC network with a maximum power of 600 hp (450 kW). The generators were driven by a water turbine supplied from the hydraulic power network. The hydraulic power network was not in competition with the electric power supply, but was seen as a supplement to it, and continued to supply power to many customer until the economic crisis of the 1930s, when the demand for pressurized water as an energy source declined. The last water engine was decommissioned in 1958.

In order to avoid excessive pressure build-up in the hydraulic power network, a release valve was fitted beside the main hall of the powerhouse. A tall water fountain, the Jet d'Eau, was ejected by the device whenever it was activated. This typically happened at the end of the day when the factories switched off their machines, making it hard to control the pressure in the system, and to adjust the supply of pressurized water to match the actual demand. The tall fountain was visible from a great distance and became a landmark in the city. When an engineering solution was found which made the fountain redundant, there was an outcry, and in 1891 it was moved to its current location in the lake, where it operated solely as a tourist attraction, although the water to create it still came from the hydraulic network.

HYDRAULIC RAM

A hydraulic ram, or hydram, is a cyclic water pump powered by hydropower. It takes in water at one "hydraulic head" (pressure) and flow rate, and outputs water at a higher hydraulic head and lower flow rate. The device uses the water hammer effect to develop pressure that allows a portion of the input water that powers the pump to be lifted to a point higher than where the water originally started. The hydraulic ram is sometimes used in remote areas, where there is both a source of low-head hydropower and a need for pumping water to a destination higher in elevation than the source. In this situation, the ram is often useful, since it requires no outside source of power other than the kinetic energy of flowing water.

A hydraulic ram that drives a fountain at the Centre for Alternative Technology.

Construction and Principle of Operation

A traditional hydraulic ram has only two moving parts, a spring or weight loaded "waste" valve sometimes known as the "clack" valve and a "delivery" check valve, making it cheap to build, easy to maintain, and very reliable.

The Papa hydraulic ram pump also has only two moving parts. However, these are not mechanical but instead rely on the elastomeric properties of the main valve and non-return valve which are low-cost, reliable and easy to replace. In addition, there is a drive pipe supplying water from an elevated source and a delivery pipe, taking a portion of the water that comes through the drive pipe to an elevation higher than the source.

Sequence of Operation

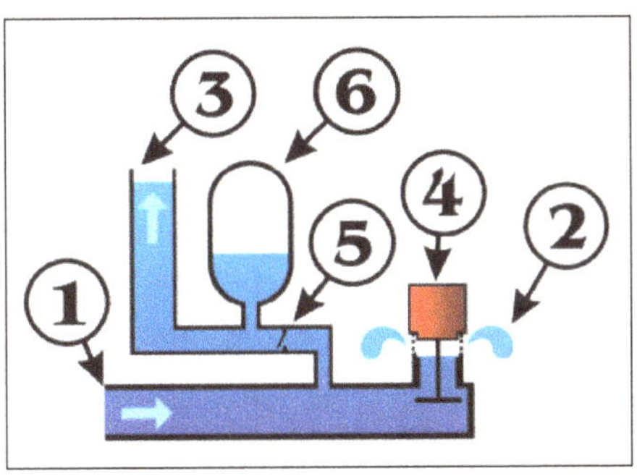

Basic components of a hydraulic ram: 1. Inlet – drive pipe 2. Free flow at waste valve 3. Outlet – delivery pipe 4. Waste valve 5. Delivery check valve 6. Pressure vessel.

A simplified hydraulic ram is shown in figure. Initially, the waste valve is open (i.e. lowered) because of its own weight, and the delivery valve is closed under the pressure caused by the water column from the outlet. The water in the inlet pipe starts to flow under the force of gravity and picks up speed and kinetic energy until the increasing drag force lifts the waste valve's weight and closes it. The momentum of the water flow in the inlet pipe against the now closed waste valve causes a water hammer that raises the pressure in the pump beyond the pressure caused by the water column pressing down from the outlet. This pressure differential now opens the delivery valve , and forces some water to flow into the delivery pipe. Because this water is being forced uphill through the delivery pipe farther than it is falling downhill from the source, the flow slows; when the flow reverses, the delivery check valve closes. Meanwhile, the water hammer from the closing of the waste valve also produces a pressure pulse which propagates back up the inlet pipe to the source where it converts to a suction pulse that propagates back down the inlet pipe. This suction pulse, with the weight or spring on the valve, pulls the waste valve back open and allows the process to begin again.

A pressure vessel containing air cushions the hydraulic pressure shock when the waste valve closes, and it also improves the pumping efficiency by allowing a more constant flow through the delivery pipe. Although the pump could in theory work without it, the efficiency would drop drastically and the pump would be subject to extraordinary stresses that could shorten its life considerably. One problem is that the pressurized air will gradually dissolve into the water until none remains. One solution to this problem is to have the air separated from the water by an elastic diaphragm (similar to an expansion tank); however, this solution can be problematic in developing countries where replacements are difficult to procure. Another solution is to have a mechanism such as a snifting valve that automatically inserts a small bubble of air when the suction pulse described above reaches the pump. Another solution is to insert an inner tube of a car or bicycle tire into the pressure vessel with some air in it and the valve closed. This tube is in effect the same as the diaphragm, but it is implemented with more widely available materials. The air in the tube cushions the shock of the water the same as the air in other configurations does.

Efficiency

A typical energy efficiency is 60%, but up to 80% is possible. This should not be confused with the volumetric efficiency, which relates the volume of water delivered to total water taken from the source. The portion of water available at the delivery pipe will be reduced by the ratio of the delivery head to the supply head. Thus if the source is 2 meters above the ram and the water is lifted to 10 meters above the ram, only 20% of the supplied water can be available, the other 80% being spilled via the waste valve. These ratios assume 100% energy efficiency. Actual water delivered will be further reduced by the energy efficiency factor. In the above example, if the energy efficiency is 70%, the water delivered will be 70% of 20%, i.e. 14%. Assuming a 2-to-1 supply head to delivery head ratio and 70% efficiency, the delivered water would be 70% of 50%, i.e. 35%. Very high ratios of delivery to supply head usually result in lowered energy efficiency. Suppliers of rams often provide tables giving expected volume ratios based on actual tests.

Drive and Delivery Pipe Design

Since both efficiency and reliable cycling depend on water hammer effects, the drive pipe design is important. It should be between 3 and 7 times longer than the vertical distance between the source and the ram. Commercial rams may have an input fitting designed to accommodate this optimum slope. The diameter of the supply pipe would normally match the diameter of the input fitting on the ram, which in turn is based on its pumping capacity. The drive pipe should be of constant diameter and material, and should be as straight as possible. Where bends are necessary, they should be smooth, large diameter curves. Even a large spiral is allowed, but elbows are to be avoided. PVC will work in some installations, but steel pipe is preferred, although much more expensive. If valves are used they should be a free flow type such as a ball valve or gate valve.

The delivery pipe is much less critical since the pressure vessel prevents water hammer effects from traveling up it. Its overall design would be determined by the allowable pressure drop based on the expected flow. Typically the pipe size will be about half that of the supply pipe, but for very long runs a larger size may be indicated. PVC pipe and any necessary valves are not a problem.

Starting Operation

A ram newly placed into operation or which has stopped cycling should start automatically if the waste valve weight or spring pressure is adjusted correctly, but it can be restarted as follows: If the waste valve is in the raised (closed) position, it must be pushed down manually into the open position and released. If the flow is sufficient, it will then cycle at least once. If it does not continue to cycle, it must be pushed down repeatedly until it cycles continuously on its own, usually after three or four manual cycles. If the ram stops with the waste valve in the down (open) position it must be lifted manually and kept up for as long as necessary for the supply pipe to fill with water and for any air bubbles to travel up the pipe to the source. This may take some time, depending on supply pipe length and diameter. Then it can be started manually by pushing it down a few times as described above. Having a valve on the delivery pipe at the ram makes starting easier. Closing the valve until the ram starts cycling, then gradually opening it to fill the delivery pipe. If opened too quickly it will stop the cycle. Once the delivery pipe is full the valve can be left open.

Common Operational Problems

Failure to deliver sufficient water may be due to improper adjustment of the waste valve, having too little air in the pressure vessel, or simply attempting to raise the water higher than the level of which the ram is capable.

The ram may be damaged by freezing in winter, or loss of air in the pressure vessel leading to excess stress on the ram parts. These failures will require welding or other repair methods and perhaps parts replacement.

It is not uncommon for an operating ram to require occasional restarts. The cycling may stop due to poor adjustment of the waste valve, or insufficient water flow at the source. Air can enter if the supply water level is not at least a few inches above the input end of the supply pipe. Other problems are blockage of the valves with debris, or improper installation, such as using a supply pipe of non uniform diameter or material, having sharp bends or a rough interior, or one that is too long or short for the drop, or is made of an insufficiently rigid material. A PVC supply pipe will work in some installations but is not as optimal as steel.

Water-powered Pump

An alternative to the hydraulic ram is the water-powered pump. It can be used if a high flow rate at high head ratio is required. A water-powered pump unit is a hydraulic turbine coupled to a water pump. The motive power needed by the pump is generated by the hydraulic turbine from the available low head water energy.

HYDRAULIC MINING

Hydraulic mining is a form of mining that uses high-pressure jets of water to dislodge rock material or move sediment. In the placer mining of gold or tin, the resulting water-sediment slurry is directed through sluice boxes to remove the gold. It is also used in mining kaolin and coal.

Hydraulic mining originated out of ancient Roman techniques that used water to excavate soft underground deposits. Its modern form, using pressurized water jets produced by a nozzle called a "monitor", came about in the 1850s during the California Gold Rush in the United States. Though successful in extracting gold-rich minerals, the widespread use of the process resulted in extensive environmental damage, such as increased flooding and erosion, and sediment blocking waterways and covering over farm fields. These problems led to its legal regulation. Hydraulic mining has since been used in various forms around the world.

A miner using a hydraulic jet to mine for gold in California.

Precursor: Ground Sluicing

Hydraulic mining had its precursor in the practice of ground sluicing, a development of which is also known as "hushing", in which surface streams of water were diverted so as to erode gold-bearing gravels. This was originally used in the Roman empire in the first centuries AD and BC, and expanded throughout the empire wherever alluvial deposits occurred. The Romans used ground sluicing to remove overburden and the gold-bearing debris in Las Médulas of Spain, and Dolaucothi in Britain. The method was also used in Elizabethan England and Wales (and rarely, Scotland) for developing lead, tin and copper mines.

Water was used on a large scale by Roman engineers in the first centuries BC and AD when the Roman empire was expanding rapidly in Europe. Using a process later known as hushing, the Romans stored a large volume of water in a reservoir immediately above the area to be mined; the water was then quickly released. The resulting wave of water removed overburden and exposed bedrock. Gold veins in the bedrock were then worked using a number of techniques, and water power was used again to remove debris. The remains at Las Médulas and in surrounding areas show badland scenery on a gigantic scale owing to hydraulicking of the rich alluvial gold deposits. Las Médulas is now a UNESCO World Heritage site. The site shows the remains of at least seven large aqueducts of up to 30 miles (48 km) in length feeding large supplies of water into the site. The gold-mining operations were described in vivid terms by Pliny the Elder in his *Naturalis Historia* published in the first century AD. Pliny was a procurator in Hispania Terraconensis in the 70s AD and must have witnessed for himself the operations. The use of hushing has been confirmed by field survey and archaeology at Dolaucothi in South Wales, the only known Roman gold mine in Britain.

California Gold Rush

Gold miners excavate an eroded bluff with jets of water at a placer mine in Dutch Flat, California sometime between 1857 and 1870.

The modern form of hydraulic mining, using jets of water directed under very high pressure through hoses and nozzles at gold-bearing upland paleogravels, was first used by Edward Matteson near Nevada City, California in 1853 during the California Gold Rush. Matteson used canvas hose which was later replaced with crinoline hose by the 1860s. In California, hydraulic mining often brought water from higher locations for long distances to holding ponds several hundred feet above the area to be mined. California hydraulic mining exploited gravel deposits, making it a form of placer mining.

Early placer miners in California discovered that the more gravel they could process, the more gold they were likely to find. Instead of working with pans, sluice boxes, long toms, and rockers, miners collaborated to find ways to process larger quantities of gravel more rapidly. Hydraulic mining became the largest-scale, and most devastating, form of placer mining. Water was redirected into an ever-narrowing channel, through a large canvas hose, and out through a giant iron nozzle, called a "monitor". The extremely high pressure stream was used to wash entire hillsides through enormous sluices.

By the early 1860s, while hydraulic mining was at its height, small-scale placer mining had largely exhausted the rich surface placers, and the mining industry turned to hard rock (called quartz mining in California) or hydraulic mining, which required larger organizations and much more capital. By the mid-1880s, it is estimated that 11 million ounces of gold (worth approximately US$7.5 billion at mid-2006 prices) had been recovered by hydraulic mining in the California Gold Rush.

Environmental Consequences

A man leans over a wooden sluice. Rocks line the outside of the wood boards that create the sluice.

While generating millions of dollars in tax revenues for the state and supporting a large population of miners in the mountains, hydraulic mining had a devastating effect on riparian natural environment and agricultural systems in California. Millions of tons of earth and water were delivered to mountain streams that fed rivers flowing into the Sacramento Valley. Once the rivers reached the relatively flat valley, the water slowed, the rivers widened, and the sediment was deposited in the floodplains and river beds causing them to rise, shift to new channels, and overflow their banks, causing major flooding, especially during the spring melt.

Cities and towns in the Sacramento Valley experienced an increasing number of devastating floods, while the rising riverbeds made navigation on the rivers increasingly difficult. Perhaps no other city experienced the boon and the bane of gold mining as much as Marysville. Situated at the confluence of the Yuba and Feather rivers, Marysville was the final "jumping off" point for miners heading to the northern foothills to seek their fortune. Steamboats from San Francisco, carrying miners and supplies, navigated up the Sacramento River, then the Feather River to Marysville where they would unload their passengers and cargo. Marysville eventually constructed a complex levee system to protect the city from floods and sediment. Hydraulic mining greatly exacerbated the problem of flooding in Marysville and shoaled the waters of the Feather River so severely that few steamboats could navigate from Sacramento to the Marysville docks. The sediment left by such efforts were reprocessed by mining dredges at the Yuba Goldfields, located near Marysville.

A similar landscape can be seen at Las Médulas in northern Spain, where Roman engineers ground-sluiced the rich gold alluvial deposits of the river Sil. Pliny the Elder mentions in his *Naturalis Historia* that Spain had encroached on the sea and local lakes as a result of ground sluicing operations.

The Malakoff Diggins, California, showing the effects of hydraulic mining on a hillside over a century later. Much of the effects of the mining was beyond the hills themselves, on the areas downstream of the water and sediment flow they produced.

Legal Consequences

Vast areas of farmland in the Sacramento Valley were deeply buried by the mining sediment. Frequently devastated by flood waters, farmers demanded an end to hydraulic mining. In the most renowned legal fight of farmers against miners, the farmers sued the hydraulic mining operations and the landmark case of *Woodruff v. North Bloomfield Mining and Gravel Company* made its way to the United States District Court in San Francisco where Judge Lorenzo Sawyer decided in favor of the farmers and banned hydraulic mining on January 7, 1884, declaring that hydraulic mining was "a public and private nuisance" and enjoining its operation in areas tributary to navigable streams and rivers. Hydraulic mining on a much smaller scale was recommenced after 1893 when the United States Congress passed the Camminetti Act which allowed such mining if sediment detention structures were constructed. This led to a number of operations above sediment catching brush dams and log crib dams. Most of the water-delivery hydraulic mining infrastructure had been destroyed by an 1891 flood, so this later stage of mining was carried on at a much smaller scale in California.

Beyond California

Although often associated with California due to its adoption and widespread use there, the technology was exported widely, to Oregon, Colorado, Montana, Arizona, Idaho, South Dakota, Alaska, British Columbia (Canada), and overseas. It was used extensively in Dahlonega, Georgia and continues to be used in developing nations, often with devastating environmental consequences. The devastation caused by this method of mining caused Edwin Carter, the "Log Cabin Naturalist", to switch from mining to collecting wildlife specimens from 1875–1900 in Breckenridge, Colorado, US.

Hydraulic mining was also used during the Australian gold rushes where it was called hydraulic sluicing. One notable location was at the Oriental Claims near Omeo in Victoria where it was used between the 1850s and early 1900s, with abundant evidence of the damage still being visible today.

The Oriental Claims near Omeo, Australia were mined between the 1850s and 1900s; hydraulic sluicing left man-made cliffs up to 30 metres (98 ft) high throughout the area.

Hydraulic mining was used extensively in the Central Otago Gold Rush that took place in the 1860s in the South Island of New Zealand, where it was also known as *sluicing*.

Starting in the 1870s, hydraulic mining became a mainstay of alluvial tin mining on the Malay Peninsula. Hydraulicking was formerly used in Polk County, Florida to mine phosphate rock.

Contemporary Usage

In addition to its use in true mining, hydraulic mining can be used as an excavation technique, principally to demolish hills. For example, the Denny Regrade in Seattle was largely accomplished by hydraulic mining.

Hydraulic mining is the principal way that kaolinite clay is mined in Cornwall and Devon, in South-West England.

Modern hydraulic mining monitor in use.

Rand Gold Fields

On the South African Rand gold fields, a gold surface tailings re-treatment facility called East Rand Gold and Uranium Company (ERGO) has been in operation since 1977. The facility uses hydraulic monitors to create slurry from older (and consequently richer) tailings sites and pumps it long distances to a concentration plant.

The facility processes nearly two million tons of tailings each month at a processing cost of below US$ 3.00/t (2013). Gold is recovered at a rate of only 0.20 g/t, but the low yield is compensated for by the extremely low cost of processing, with no risky or expensive mining or milling required for recovery.

The resulting slimes are pumped further away from the built-up areas permitting the economic development of land close to commercially valuable areas and previously covered by the tailings. The historic yellow-coloured mine dumps around Johannesburg are now almost a rarity, seen only in older photographs.

Uranium and pyrite (for sulfuric acid production) are also available for recovery from the process stream as co-products under suitable economic conditions.

Underground Hydraulic Mining

High-pressure water jets have also been used in the underground mining of coal, to break up the coal seam and wash the resulting coal slurry toward a collection point. The high-pressure water nozzle is referred to as the 'hydro monitor'.

References

- Dam-engineering, technology: britannica.com, Retrieved 8 June, 2019
- "Micro hydro in the fight against poverty". Tve.org. Tve/itdg. November 2004. Archived from the original on 30 july 2007. Retrieved 14 january 2017
- How-hydropower-works: wvic.com, Retrieved 5 March, 2019
- "Unido, icshp launch small hydropower knowledge sharing portal". Sustainable energy policy and practice. Retrieved 29 april 2015
- Hydroelectric-power-plant-layout: electricaleasy.com, Retrieved 1 August, 2019
- "Micro hydro project". Technologies for conservation & development project. Archived from the original on 31 july 2009. Retrieved 6 september2010
- Rehman, shafiqur; al-hadhrami, luai; alam, md (30 april 2015). "pumped hydro energy storage system: a technological review". Renewable and sustainable energy reviews. 44: 586–598. Doi:10.1016/j.rser.2014.12.040 – via researchgate
- Dwivedi, a.k. Raja, amit prakash srivastava, manish (2006). Power plant engineering. New delhi: new age international. P. 354. Isbn 81-224-1831-7

4

Hydroelectricity Equipments

The main equipment used for producing hydroelectricity are water Turbines and water wheels. Some of the common types of water Turbines are Francis Turbine, Kaplan Turbine, Gorlov Helical Turbine, Pelton Turbine, Turgo Turbine and Cross-flow Turbine. This chapter discusses these hydroelectricity equipment in detail.

WATER TURBINE

The Water Turbine, also known as a Hydro Turbine, is a fairly simple machine that produces a rotary turning action at a medium to high rotational speed. Also, water turbines can be used as part of a home hydro electricity system by installing an electrical generator. Unlike a water pump which is mechanically driven by an electric motor or wind turbine and uses suction to pump the water through it, a typical water turbine design uses nozzles and differential water pressure to produce a mechanical rotation and output.

Typical Water Turbine Design.

A waterwheel is a simple but large circular wooden or metallic wheel with buckets attached around its periphery that slowly turns as the flow of water pours over or underneath it producing lots of mechanical torque to drive auxiliary machinery. A water turbine on the other hand, is a much smaller, lightweight cast iron or steel machine in which the kinetic energy of the water is converted into mechanical energy through the use of correctly placed pressure nozzles.

The water turbine is the heart of any hydro power plant. It consists of a number of metal or plastic blades fitted to a central rotating shaft or plate. Water flowing through the casing of the enclosed turbine, strikes the blades of the turbine producing torque and making the shaft rotate due to the velocity and pressure of the water. As the water pushes against the turbine blades, its velocity and pressure reduces (energy is lost) as it rotates the turbine shaft.

However, as a typical "water turbine design" has lots more blades attached to this central shaft, this rotation causes the next blade behind it to come into contact with the entering water pressure causing the turbine to rotate some more and so on.

As the turbine continues to rotate, the water becomes trapped in between the turbines blades and is pushed along by the rotational movement of the turbine. At some point along the rotational angle of the turbine blades, the water encounters an opening in the casing, usually located at the centre, which allows the water to exit and return back to the river or stream from where it originally came.

This waste water can pass through the body of a water turbine using different flow paths and based upon the actual path of water flow through the turbine, modern water turbine designs can be categorised as being either:

- Axial Flow Turbines: The water flow path through an axial flow water turbine design is parallel to the axis of rotation as it enters the turbines wheel from the side.
- Radial Flow Turbines: The water flows path through a radial flow water turbine design is perpendicular to the axis of rotation of the turbine as it enters the turbines wheel from above.

There are many different Water Turbine Designs in use today, with each type having its own advantages and disadvantages depending upon their operational requirements. The selection of the correct type of water turbine design is very important for the success of any small or large scale hydro power system.

The optimum mechanical power output from the turbines rotating shaft relies on a specific combination of head height, water volume and water pressure entering the turbines blades, which can only be achieved by selecting the correct type of water turbine for a given installation. Then based upon the actual change in the waters pressure as it used by the turbines wheel, modern water turbine designs can be categorised as either:

- Reaction Turbine Design – in this type of water turbine design, the turbine blades are totally submersed in the flow of the water and are enclosed within a pressurised casing. A reaction turbine is powered mainly by the change in pressure, called a "pressure drop" across the casings body as this reduction in water pressure and velocity releases energy causing a reaction (hence the name) by moving the turbines blades. The flow of water through a reaction turbine may be reversed due to the angle of the internal blades, so a reaction turbine can also be used to pump water and vice versa.

- Impulse Turbine Design – in this type of water turbine design, the water flow hits the turbine blades from one or more jets of water known as nozzles. These nozzles convert the pressurised low velocity water into a high speed jet of water aimed directly at the turbines curved spoon or bucket shaped blades generating maximum force on the blades. The mechanical power output from an impulse turbine is derived from the kinetic energy of the water flow. An impulse turbine operates in a fully-open or semi-open casing therefore there is no pressure drop across and impulse turbine and due to its open nozzle jet design it can not be reversed.

Types of Water Turbine Design

As well as the type of operation of a water turbine, water turbines are often referred to by the design, structure and layout of their blades inside the actual turbine body. Names such as: "Pelton Turbines", "Turgo Turbines" or "Kaplan Turbines" all refer to the name of the designer and/or inventor of that particular water turbine design. Following are some of the more common types of water turbine designs. But before we look at the different types of water turbines available, we first need to understand some of the "jargon" associated with water turbines.

The word Head is the term used in hydro energy applications to define the vertical distance the water must fall from a higher elevation to a lower point in order to release its stored potential energy. The difference between these two elevations (the higher minus the lower) is called the "head" of the system. Generally, in hydro energy and hydro-electric power systems, a low-head means a vertical distance of less than 100 feet (approx 30 metres), a medium-head means a vertical distance of between 100 to 500 feet (30-to-150 metres), while a high-head means a vertical distance greater than 500 feet, (150 metres).

The amount of head available in hydro energy system determines the amount of power available from the water as Power is proportional to Head x Flow. The flow of water through the turbine is normally expressed in litres per second, (ltr/s) or cubic metres per second (m^3/s) and refers to the quantity or volume of water used by the turbine to turn the shaft.

Francis Turbine

Francis Turbine is a combination of both impulse and reaction turbine, where the blades rotate using both reaction and impulse force of water flowing through them producing electricity more efficiently. Francis turbine is used for the production of electricity in hydro power stations.

Majorly there are 2 turbines flow patterns on which they work, namely radial and axial flow concepts. An American civil engineer by name, James B. Francis in Lowell, Massachusetts comes up with an idea of combining both impulse and reaction turbine where water enters the turbine radically and exits axially.

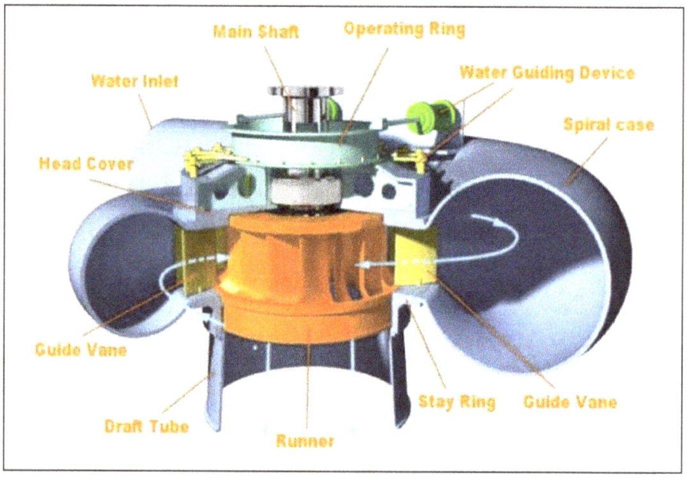

Components of Francis Turbine.

The main reason of higher efficiency of Francis turbine lies in the design of blades, these blades rotate using both reaction and impulse force of water flowing through them. Due the use of this type of turbines the main problem faced due to the water head availability is eliminated as the turbine uses both the kinetic and potential energy to produce power.

Main Components of Francis Turbine

The major components of Francis turbine are:

Spiral Casing

Spiral casing is the inlet medium of water to the turbine. The water flowing from the reservoir or dam is made to pass through this pipe with high pressure. The blades of the turbines are circularly placed, which mean the water striking the turbines blades should flow in the circular axis for efficient striking. So the spiral casing is used, but due to circular movement of the water, it looses its pressure.

To maintain the same pressure the diameter of the casing is gradually reduced, so as to maintain the pressure uniform, thus uniform momentum or velocity striking the runner blades.

Stay Vanes

Stay vanes and guide vanes guides the water to the runner blades. Stay vanes remain stationary at their position and reduces the swirling of water due to radial flow, as it enters the runner blades. Thus making turbine more efficient.

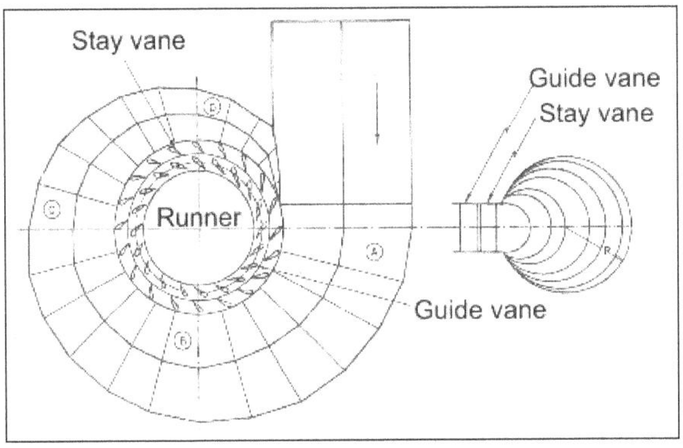

Stay Vanes and Guide Vanes of Francis Turbine.

Guide Vanes

Guide vanes are not stationary, they change their angle as per the requirement to control the angle of striking of water to turbine blades to increase the efficiency. They also regulate the flow rate of water into the runner blades thus controlling the power output of a turbine according to the load on the turbine.

Runner Blades

The performance and efficiency of the turbine is dependent on the design of the runner blades. In a Francis turbine, runner blades are divided into 2 parts. The lower half is made in the shape of small bucket so that it uses the impulse action of water to rotate the turbine.

The upper part of the blades use the reaction force of water flowing through it. These two forces together makes the runner to rotate.

Draft Tube

The pressure at the exit of the runner of Reaction Turbine is generally less than atmospheric pressure. The water at exit cannot be directly discharged to the tail race. A tube or pipe of gradually increasing area is used for discharging water from the exit of turbine to the tail race.

This tube of increasing area is called Draft Tube. One end of the tube is connected to the outlet of runner while the other end is sub-merged below the level of water in the tail-race.

Working

The water is allowed to enter the spiral casing of the turbine, which lead the water through the stay vanes and guide vanes. The spiral case is kept in decreasing diameter so as to maintain the flow pressure.

The stay vanes being stationary at their place, removes the swirls from the water, which are generated due to flow through spiral casing and tries it to make the flow of water more linear to be deflected by adjustable guide vanes.

The angle of guide vanes decides the angle of attack of water at the runner blades thus make sure the output of the turbine. The runner blades are stationary and can-not pitch or change their angle so it's all about the guide vanes which controls the power output of a turbine.

The performance and efficiency of the turbine is dependent on the design of the runner blades. In a Francis turbine, runner blades are divided into 2 parts. The lower half is made in the shape of small bucket so that it uses the impulse action of water to rotate the turbine.

The upper part of the blades use the reaction force of water flowing through it. Thus runner blades make use of both pressure energy and kinetic energy of water and rotates the runner in most efficient way.

The water coming out of runner blades would lack both the kinetic energy and pressure energy, so we use the draft tube to recover the pressure as it advances towards tail race, but still we cannot recover the pressure to that extent that we can stop air to enter into the runner housing thus causing cavitation.

External Francis Turbine attached to a dynamo.

Applications of Francis Turbine

- Francis turbine is the most widely used turbine in hydro-power plants to generate electricity.
- Mixed flow turbine is also used in irrigation water pumping sets to pump water from ground for irrigation.

- It is efficient over a wide range of water head and flow rate.
- It is most efficient hydro-turbine we have till date.

Kaplan Turbine

The Kaplan turbine is a propeller-type water turbine which has adjustable blades. It was developed in 1913 by Austrian professor Viktor Kaplan, who combined automatically adjusted propeller blades with automatically adjusted wicket gates to achieve efficiency over a wide range of flow and water level.

The Kaplan turbine was an evolution of the Francis turbine. Its invention allowed efficient power production in low-head applications which was not possible with Francis turbines. The head ranges from 10–70 metres and the output ranges from 5 to 200 MW. Runner diameters are between 2 and 11 metres. Turbines rotate at a constant rate, which varies from facility to facility. That rate ranges from as low as 54.5 rpm (Albeni Falls Dam) to 450 rpm.

Kaplan turbines are now widely used throughout the world in high-flow, low-head power production.

A Bonneville Dam Kaplan turbine after 61 years of service.

On this Kaplan runner the pivots at the base of the blade are visible; these allow the angle of the blades to be changed while running. The hub contains hydraulic cylinders for adjusting the angle.

Development

Viktor Kaplan living in Brno, Austria-Hungary, obtained his first patent for an adjustable blade propeller turbine in 1912. But the development of a commercially successful machine would take another decade. Kaplan struggled with cavitation problems, and in 1922 abandoned his research for health reasons.

In 1919 Kaplan installed a demonstration unit at Poděbrady, Czechoslovakia. In 1922 Voith introduced an 1100 HP (about 800 kW) Kaplan turbine for use mainly on rivers. In 1924 an 8 MW unit went on line at Lilla Edet, Sweden. This marked the commercial success and widespread acceptance of Kaplan turbines.

Theory of Operation

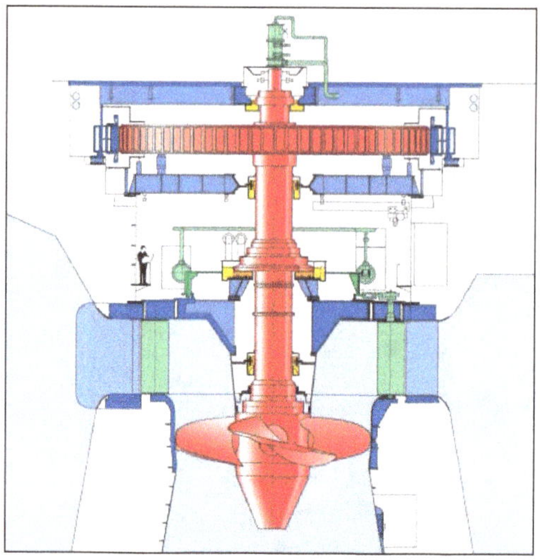

Vertical Kaplan Turbine.

The Kaplan turbine is an inward flow reaction turbine, which means that the working fluid changes pressure as it moves through the turbine and gives up its energy. Power is recovered from both the hydrostatic head and from the kinetic energy of the flowing water. The design combines features of radial and axial turbines.

The inlet is a scroll-shaped tube that wraps around the turbine's wicket gate. Water is directed tangentially through the wicket gate and spirals on to a propeller shaped runner, causing it to spin.

The outlet is a specially shaped draft tube that helps decelerate the water and recover kinetic energy.

The turbine does not need to be at the lowest point of water flow as long as the draft tube remains full of water. A higher turbine location, however, increases the suction that is imparted on the turbine blades by the draft tube. The resulting pressure drop may lead to cavitation.

Variable geometry of the wicket gate and turbine blades allow efficient operation for a range of flow conditions. Kaplan turbine efficiencies are typically over 90%, but may be lower in very low head applications.

Current areas of research include computational fluid dynamics (CFD) driven efficiency improvements and new designs that raise survival rates of fish passing through.

Because the propeller blades are rotated on high-pressure hydraulic oil bearings, a critical element of Kaplan design is to maintain a positive seal to prevent emission of oil into the waterway. Discharge of oil into rivers is not desirable because of the waste of resources and resulting ecological damage.

Applications

Viktor Kaplan Turbine Technisches Museum Wien.

Kaplan turbines are widely used throughout the world for electrical power production. They cover the lowest head hydro sites and are especially suited for high flow conditions.

Inexpensive micro turbines on the Kaplan turbine model are manufactured for individual power production designed for 3 m of head which can work with as little as 0.3 m of head at a highly reduced performance provided sufficient water flow.

Large Kaplan turbines are individually designed for each site to operate at the highest possible efficiency, typically over 90%. They are very expensive to design, manufacture and install, but operate for decades. They have recently found a new home in offshore wave energy generation.

Variations

The Kaplan turbine is the most widely used of the propeller-type turbines, but several other variations exist:

- Propeller turbines have non-adjustable propeller vanes. They are used where the range of flow/power is not large. Commercial products exist for producing several hundred watts from only a few feet of head. Larger propeller turbines produce more than 100 MW. At the La Grande-1 generating station in northern Quebec, 12 propeller turbines generate 1368 MW.

- Bulb or tubular turbines are designed into the water delivery tube. A large bulb is centered in the water pipe which holds the generator, wicket gate and runner. Tubular turbines are a fully axial design, whereas Kaplan turbines have a radial wicket gate.
- Pit turbines are bulb turbines with a gear box. This allows for a smaller generator and bulb.
- Straflo turbines are axial turbines with the generator outside of the water channel, connected to the periphery of the runner.
- S-turbines eliminate the need for a bulb housing by placing the generator outside of the water channel. This is accomplished with a jog in the water channel and a shaft connecting the runner and generator.
- The VLH turbine is an open flow, very low head "kaplan" turbine slanted at an angle to the water flow. It has a large diameter >3.55 m, is low speed using a directly connected shaft mounted permanent magnet alternator with electronic power regulation and is very fish friendly (<5% mortality).
- The DIVE-Turbine is a vertical propeller turbine with double regulation by wicket gates and speed variation. It covers a range of application up to 4 MW with efficiencies comparable to standard Kaplan-Turbines. Due to the propeller design with fixed blades it is considered a fish friendly turbine.

DIVE-Turbine, a propeller turbine version, during installation on site.

- Tyson turbines are a fixed propeller turbine designed to be immersed in a fast flowing river, either permanently anchored in the river bed, or attached to a boat or barge.

Gorlov Helical Turbine

The Gorlov helical turbine (GHT) is a water turbine evolved from the Darrieus turbine design by altering it to have helical blades/foils. It was patented in a series of patents from September 19, 1995 to July 3, 2001 and won 2001 ASME Thomas A. Edison Patent Award. GHT was invented by Alexander M. Gorlov, professor of Northeastern University.

The physical principles of the GHT work are the same as for its main prototype, the Darrieus turbine, and for the family of similar vertical axis wind turbines which includes also Turby wind turbine, aerotecture turbine, Quietrevolution wind turbine, etc. GHT, Turby and Quietrevolution solved pulsatory torque issues by using the helical twist of the blades. The original inventor of the helical turbine was Ulrich Stampa.

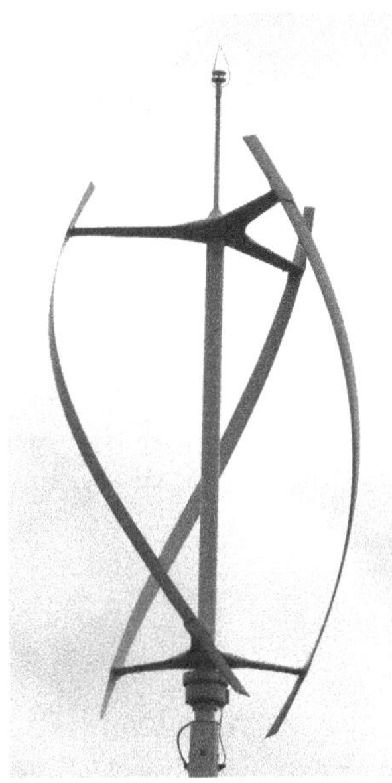

quietrevolution QR5 wind turbine.

Fluid Performance

The term "foil" is used to describe the shape of the blade cross-section at a given point, with no distinction for the type of fluid, (thus referring to either an "airfoil" or "hydrofoil"). In the helical design, the blades curve around the axis, which has the effect of evenly distributing the foil sections throughout the rotation cycle, so there is always a foil section at every possible angle of attack. In this way, the sum of the lift and drag forces on each blade do not change abruptly with rotation angle. The turbine generates a smoother torque curve, so there is much less vibration and noise than in the Darrieus design. It also minimizes peak stresses in the structure and materials, and facilitates self-starting of the turbine. In testing environments the GHT has been observed to have up to 35% efficiency in energy capture reported by several groups. "Among the other vertical-axis turbine systems, the Davis Hydro turbine, the EnCurrent turbine, and the

Gorlov Helical turbine have all undergone scale testing at laboratory or sea. Overall, these technologies represent the current norm of tidal current development."

Turbine Axis-orientation

The main difference between the Gorlov helical turbine and conventional turbines is the orientation of the axis in relation to current flow. The GHT is a vertical-axis turbine which means the axis is positioned perpendicular to current flow, whereas traditional turbines are horizontal-axis turbines which means the axis is positioned parallel to the flow of the current. Fluid flows, such as wind, will naturally change direction, however they will still remain parallel to the ground. So in all vertical-axis turbines, the flow remains perpendicular to the axis, regardless of the flow direction, and the turbines always rotate in the same direction. This is one of the main advantages of vertical-axis turbines.

If the direction of the water flow is fixed, then the Gorlov turbine axis could be vertical or horizontal, the only requirement is orthogonality to the flow.

Airfoil/Hydrofoil

The GHT operates under a lift-based concept. The foil sections on the GHT are symmetrical, both top-to-bottom and also from the leading-to-trailing edge. The GHT can actually spin equally well in either direction. The GHT works under the same principle as the Darrieus turbine; that is, it relies upon the movement of the foils in order to change the apparent direction of the flow relative to the foils, and thus change the (apparent) "angle of attack" of the foil.

Environmental Issues

A GHT is proposed for low-head micro hydro installations, when construction of a dam is undesirable. The GHT is an example of damless hydro technology. The technology may potentially offer cost and environmental benefits over dam-based micro-hydro systems.

Some advantages of damless hydro are that it eliminates the potential for failure of a dam, which improves public safety. It also eliminates the initial cost of dam engineering, construction and maintenance, reduces the environmental and ecological complications, and potentially simplifies the regulatory issues put into law specifically to mitigate the problems with dams.

In general, a major ecological issue with hydropower installations is their actual and perceived risk to aquatic life. It is claimed that a GHT spins slowly enough that fish can see it soon enough to swim around it. From preliminary tests in 2001, it was claimed that if a fish swims between the slowly moving turbine blades, the fish will not be harmed. Also it would be difficult for a fish to become lodged or stuck in the turbine, because the open spaces between the blades are larger than even the largest fish living in a small river. A fish also would not be tumbled around in a vortex, because the GHT does not create a lot of turbulence, so small objects would be harmlessly swept through with the current.

Working

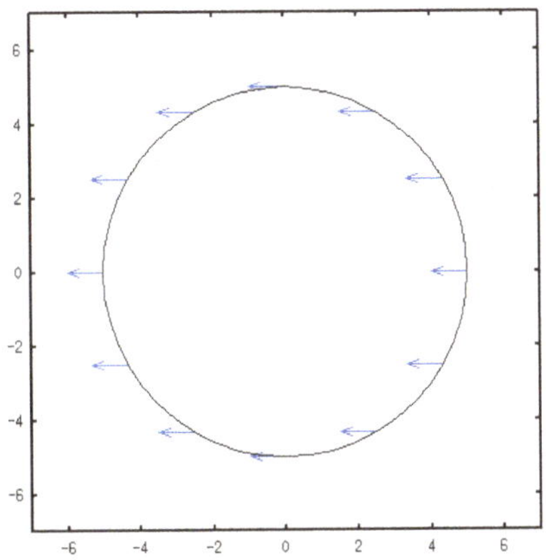

Current flow to the left.

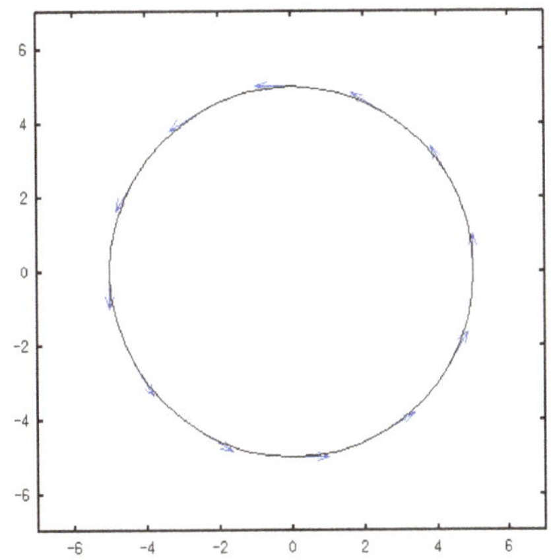

Induced flow component created by clockwise rotation of turbine.

In this example the direction of the fluid flow is to the left. As the turbine rotates, in this case in a clockwise direction, the motion of the foil through the fluid changes the apparent velocity and angle of attack (speed and direction) of the fluid with respect to the frame of reference of the foil. The combined effect of these two flow components (i.e. the vector sum), yields the net total "Apparent flow velocity" as shown in the next figure.

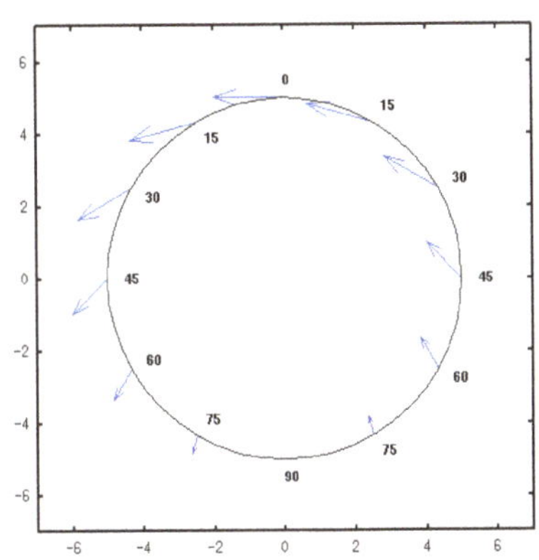

Apparent flow velocity of turbine blade, and angle made with current flow over ground (in degrees).

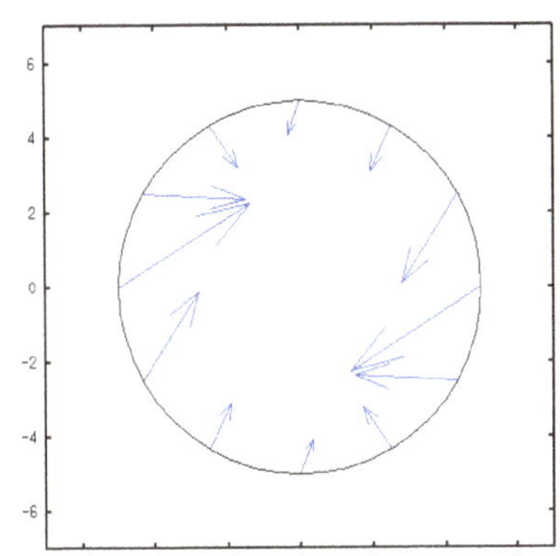

Net force vectors.

The action of this apparent flow on each foil section generates both a lift and drag force, the sum of which is shown in the figure above titled "Net force vectors". Each of these net force vectors can be split into two orthogonal vectors: a radial component and a tangential component, shown here as "Normal force" and "Axial force" respectively. The normal forces are opposed by the rigidity of the turbine structure and do not impart any rotational force or energy to the turbine. The remaining force component propels the turbine in the clockwise direction, and it is from this torque that energy can be harvested.

With regards to the figure above left “Apparent flow velocity”, Lucid Energy Technologies, rights holder to the patent to the Gorlov Helical Turbine, notes that this diagram, with no apparent velocity at an azimuth angle of 180 degrees (blade at its point in rotation where it is instantaneously moving in downstream direction), may be subject to misinterpretation. This is because a zero apparently flow velocity could occur only at a tip speed ratio of unity (i.e. TSR=1, where the current flow induced by rotation equals the current flow). The GHT generally operates at a TSR substantially greater than unity.

(The diagrams “Net Force Vectors” and “Normal Force Vectors” are partially incorrect. The downwind segments should show the vectors outside the circles. Otherwise there would be no net sideways loading on the turbine.) M Koester 2015.

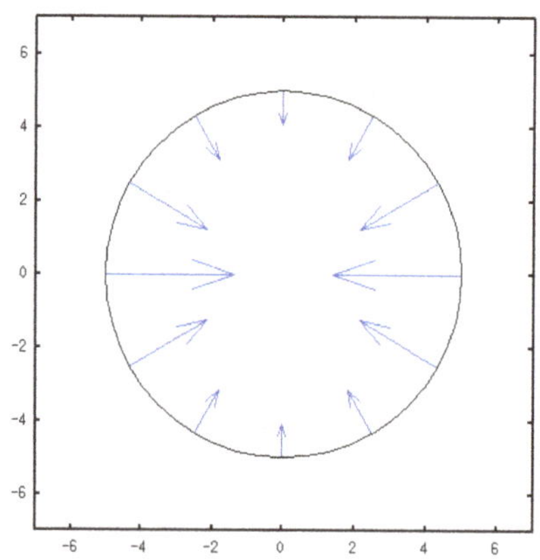

Normal force vectors.

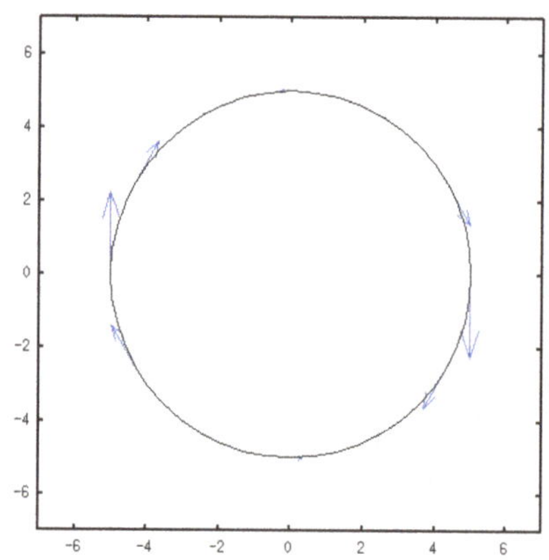

Axial force vectors.

Commercial Use

Helical turbines in water stream generate mechanical power independent on direction of the water flow. Then electric generators assembled upon the common shaft transfer the power into electricity for the commercial use.

Chain of horizontal Gorlov turbines being installed in Cobscook Bay, Maine, USA.

Chain of horizontal Gorlov turbines. TideGen by Ocean Renewable Power Company – possibility for shallow waters.

Tidal power station with Gorlov helical turbines before deployment in the ocean. USA, Cobscook Bay, Maine, September, 2012.

Chain of horizontal Gorlov turbines being installed in South Korea – general view.

Chain of horizontal Gorlov turbines being installed in South Korea – close view.

Gorlov helical turbines in South Korea, 1997–1998. Installation in shallow waters.

Pelton Turbine

A Pelton turbine or Pelton wheel is a type of turbine used frequently in hydroelectric plants. These turbines are generally used for sites with heads greater than 300 meters. This type of turbine was created during the gold rush in 1880 by Lester Pelton.

A Pelton turbine.

When used for generating electricity, there is usually a water reservoir located at some height above the Pelton turbine. The water then flows through the penstock to specialized nozzles that introduce pressurized water to the turbine. To prevent irregularities in pressure, the penstock is fitted with a surge tank that absorbs sudden fluctuations in water that could alter the pressure. Unlike other types of turbines which are reaction turbines, the Pelton turbine is known as an impulse turbine. This simply means that instead of moving as a result of a reaction force, water creates some impulse on the turbine to get it to move.

Design

The Pelton turbine has a fairly simplistic design. A large circular disk is mounted on some sort of rotating shaft known as a rotor. Mounted on this circular disk are cup shaped blades known as buckets evenly spaced around the entire wheel. Generally, the buckets are arranged in pairs around the rim. Then nozzles are arranged the wheel and serve the purpose of introducing water to the turbine. Jets of water emerge from these nozzles, tangential to the wheel of the turbine. This causes the turbine to spin as a result of the impact of the water jets on the buckets.

Operation

The operation of a Pelton turbine is fairly simple. In this type of turbine, high speed jets of water emerge from the nozzles that surround the turbine. These nozzles are arranged so the water jet will hit the buckets at splitters, the center of the bucket where the water jet is divided into two streams. The two separate streams then flow along the inner curve of the bucket and leave in the opposite direction that it came in. This change in momentum of the water creates an impulse on the blades of the turbine, generating torque and rotation in the turbine.

The high speed water jets are created by pushing high pressure water (such as water falling from high heads) through nozzles at atmospheric pressure. The maximum output is obtained from a Pelton turbine when the impulse obtained by the blades is maximum, meaning that the water stream is deflected exactly opposite to the direction at which it strikes the buckets at. As well, the efficiency of these wheels is highest when the speed of the movement of the cups is half of the speed of the water jet.

Turgo Turbine

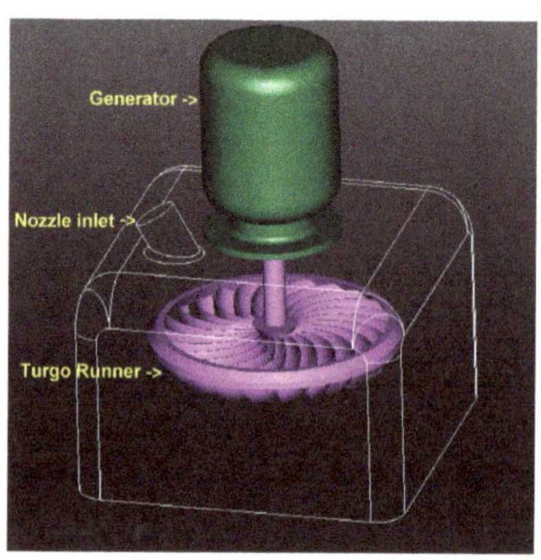

Turgo turbine and generator.

The Turgo turbine is an impulse water turbine designed for medium head applications. Operational Turgo Turbines achieve efficiencies of about 87%. In factory and lab tests Turgo Turbines perform with efficiencies of up to 90%. It works with net heads between 15 and 300 m.

Designed in 1919 by Gilkes as a modification of the Pelton wheel, the Turgo has some advantages over Francis and Pelton designs for certain applications.

First, the runner is less expensive to make than a Pelton wheel. Second, it doesn't need an airtight housing like the Francis. Third, it has higher specific speed and can handle a greater flow than the same diameter Pelton wheel, leading to reduced generator and installation cost.

Turgos operate in a head range where the Francis and Pelton overlap. While many large Turgo installations exist, they are also popular for small hydro where low cost is very important.

Like all turbines with nozzles, blockage by debris must be prevented for effective operation.

At Milford sound, New Zealand.

Theory of Operation

The Turgo turbine is an impulse type turbine; water does not change pressure as it moves through the turbine blades. The water's potential energy is converted to kinetic energy with a nozzle. The high speed water jet is then directed on the turbine blades which deflect and reverse the flow. The resulting impulse spins the turbine runner, imparting energy to the turbine shaft. Water exits with very little energy. Turgo runners are extremely efficient.

A Turgo runner looks like a Pelton runner split in half. For the same power, the Turgo runner is one half the diameter of the Pelton runner, and so twice the specific speed. The Turgo can handle a greater water flow than the Pelton because exiting water doesn't interfere with adjacent buckets.

The specific speed of Turgo runners is between the Francis and Pelton. Single or multiple nozzles can be used. Increasing the number of jets increases the specific speed of the runner by the square root of the number of jets (four jets yield twice the specific speed of one jet on the same turbine).

Cross-flow Turbine

A cross-flow turbine, Bánki-Michell turbine, or Ossberger turbine is a water turbine developed by the Australian Anthony Michell, the Hungarian Donát Bánki and the German Fritz Ossberger. Michell obtained patents for his turbine design in 1903, and the manufacturing company Weymouth made it for many years. Ossberger's first patent was granted in 1933 and he manufactured this turbine as a standard product. Today, the company founded by Ossberger is the leading manufacturer of this type of turbine.

Unlike most water turbines, which have axial or radial flows, in a cross-flow turbine the water passes through the turbine transversely, or across the turbine blades. As with a water wheel, the water is admitted at the turbine's edge. After passing to the inside of the runner, it leaves on the opposite side, going outward. Passing through the runner twice provides additional efficiency. When the water leaves the runner, it also helps clean it of small debris and pollution. The cross-flow turbine is a low-speed machine that is well suited for locations with a low head but high flow.

Although the illustration shows one nozzle for simplicity, most practical cross-flow turbines have two, arranged so that the water flows do not interfere.

Cross-flow turbines are often constructed as two turbines of different capacity that share the same shaft. The turbine wheels are the same diameter, but different lengths to handle different volumes at the same pressure. The subdivided wheels are usually built with volumes in ratios of 1:2. The subdivided regulating unit, the guide vane system in the turbine's upstream section, provides flexible operation, with 33, 66 or 100% output, depending on the flow. Low operating costs are obtained with the turbine's relatively simple construction.

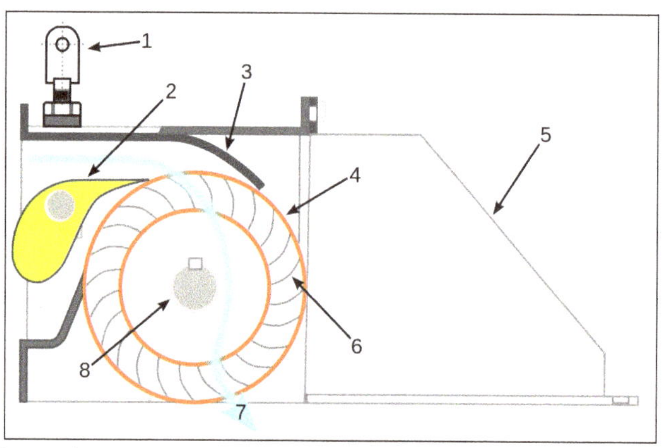

Diagram of a Cross-flow turbine (1) — air-venting valve (2) — distributor (3) — turbine casing (all thick grey) (4) — runner (5) — removable rear casing (6) — blades (7) — water flow (8) — shaft.

Details of Design

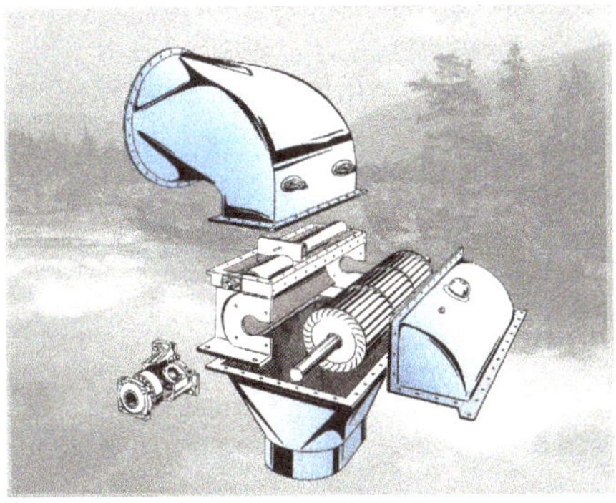

Ossberger turbine section.

The turbine consists of a cylindrical water wheel or runner with a horizontal shaft, composed of numerous blades (up to 37), arranged radially and tangentially. The blade's edges are sharpened to reduce resistance to the flow of water. A blade is made in a part-circular cross-section (pipe cut over its whole length). The ends of the blades are welded to disks to form a cage like a hamster cage and are sometimes called "squirrel cage turbines"; instead of the bars, the turbine has the trough-shaped steel blades.

The water flows first from the outside of the turbine to its inside. The regulating unit, shaped like a vane or tongue, varies the cross-section of the flow. The water jet is directed towards the cylindrical runner by nozzle. The water enters the runner at an angle of about 45/120 degrees,transmitting some of the water's kinetic energy to the active cylindrical blades.

Ossberger turbine runner.

The regulating device controls the flow based on the power needed, and the available water. The ratio is that (0–100%) of the water is admitted to 0-100%×30/4 blades. Water admission to the two nozzles is throttled by two shaped guide vanes. These divide and direct the flow so that the water enters the runner smoothly for any width of opening. The guide vanes should seal to the edges of the turbine casing so that when the water is low, they can shut off the water supply. The guide vanes therefore act as the valves between the penstock and turbine. Both guide vanes can be set by control levers, to which an automatic or manual control may be connected.

The turbine geometry (nozzle-runner-shaft) assures that the water jet is effective. The water acts on the runner twice, but most of the power is transferred on the first pass, when the water enters the runner. Only ⅓ of the power is transferred to the runner when the water is leaving the turbine.

The water flows through the blade channels in two directions: outside to inside, and inside to outside. Most turbines are run with two jets, arranged so two water jets in the runner will not affect each other. It is, however, essential that the turbine, head and turbine speed are harmonised.

The cross-flow turbine is of the impulse type, so the pressure remains constant at the runner.

Advantages

The peak efficiency of a cross-flow turbine is somewhat less than a Kaplan, Francis or Pelton turbine. However, the cross-flow turbine has a flat efficiency curve under varying load. With a split runner and turbine chamber, the turbine maintains its efficiency while the flow and load vary from 1/6 to the maximum.

Since it has a low price, and good regulation, cross-flow turbines are mostly used in mini and micro hydropower units of less than two thousand kW and with heads less than 200 m.

Particularly with small run-of-the-river plants, the flat efficiency curve yields better annual performance than other turbine systems, as small rivers' water is usually lower in some months. The efficiency of a turbine determines whether electricity is produced during the periods when rivers have low flows. If the turbines used have high peak efficiencies, but behave poorly at partial load, less annual performance is obtained than with turbines that have a flat efficiency curve.

Due to its excellent behaviour with partial loads, the cross-flow turbine is well-suited to unattended electricity production. Its simple construction makes it easier to maintain than other turbine types; only two bearings must be maintained, and there are only three rotating elements. The mechanical system is simple, so repairs can be performed by local mechanics.

Another advantage is that it can often clean itself. As the water leaves the runner, leaves, grass etc. will not remain in the runner, preventing losses. Therefore, although the turbine's efficiency is somewhat lower, it is more reliable than other types. No runner cleaning is normally necessary, e.g. by flow inversion or variations of the speed. Other turbine types are clogged more easily, and consequently face power losses despite higher nominal efficiencies.

WATER WHEEL

A water wheel is a machine for converting the energy of flowing or falling water into useful forms of power, often in a watermill. A water wheel consists of a wheel (usually constructed from wood or metal), with a number of blades or buckets arranged on the outside rim forming the driving surface.

Water wheels were still in commercial use well into the 20th century but they are no longer in common use. Uses included milling flour in gristmills, grinding wood into pulp for papermaking, hammering wrought iron, machining, ore crushing and pounding fiber for use in the manufacture of cloth.

Some water wheels are fed by water from a mill pond, which is formed when a flowing stream is dammed. A channel for the water flowing to or from a water wheel is called a mill race. The race bringing water from the mill pond to the water wheel is a headrace; the one carrying water after it has left the wheel is commonly referred to as a tailrace.

In the mid to late 18th century John Smeaton's scientific investigation of the water wheel led to significant increases in efficiency supplying much needed power for the Industrial Revolution.

Water wheels began being displaced by the smaller, less expensive and more efficient turbine, developed by Benoît Fourneyron, beginning with his first model in 1827. Turbines are capable of handling high *heads*, or elevations, that exceed the capability of practical-sized waterwheels.

The main difficulty of water wheels is their dependence on flowing water, which limits where they can be located. Modern hydroelectric dams can be viewed as the descendants of the water wheel, as they too take advantage of the movement of water downhill.

An overshot waterwheel standing 42 ft (13 m) high powers the Old Mill at Berry College in Rome, Georgia, United States.

Water wheel powering a mine hoist in *De re metallica* (1566).

Types

Water wheels come in two basic designs:

- a horizontal wheel with a vertical axle; or
- a vertical wheel with a horizontal axle.

The latter can be subdivided according to where the water hits the wheel into backshot (pitch-back) overshot, breastshot, undershot, and stream-wheels. The term undershot can refer to any wheel where the water passes under the wheel but it usually implies that the water entry is low on the wheel.

Most water wheels in the United Kingdom and the United States are (or were) vertical wheels rotating about a horizontal axle, but in the Scottish highlands and parts of Southern Europe mills often had a horizontal wheel (with a vertical axle).

Types

Vertical axis also known as tub or Norse mills.

- Horizontal wheel with a vertical axis
- A jet of water strikes blades mounted on the axle
- Driving surfaces – blades
- Water – low volume, high head
- Efficiency – poor

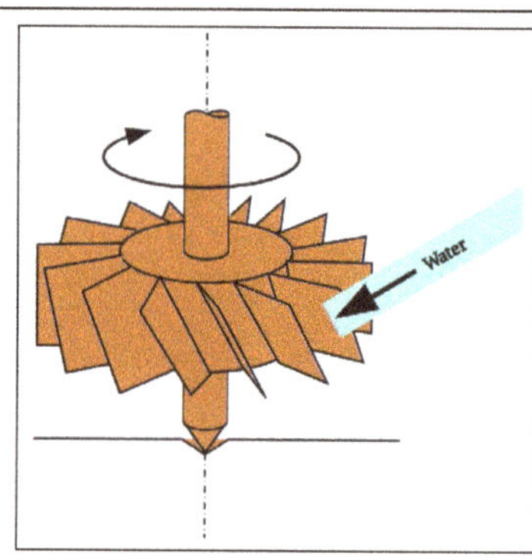

Diagram of vertical axis water mill.

Stream (also known as free surface). Ship wheels are a type of stream wheel. • Vertical wheel with horizontal axle • The bottom of the wheel is placed into flowing water • Driving surfaces – blades – flat prior to 18th century, curved thereafter • Water – very large volume, no head • Efficiency – about 20% prior to 18th century and later 50 to 60%	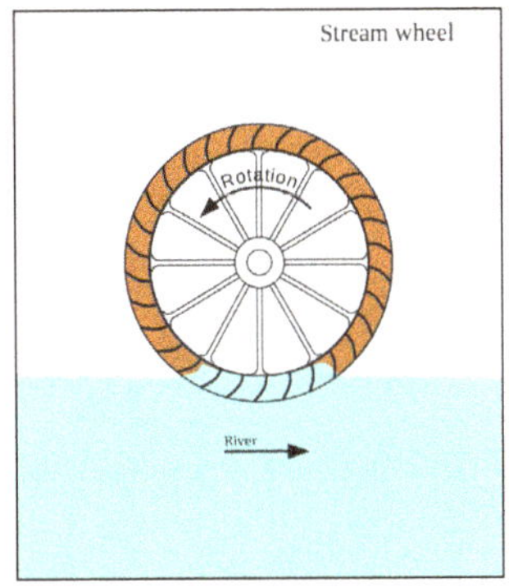 Diagram of stream shot waterwheel.
Undershot • Vertical wheel with horizontal axle • The water hits the wheel low down, typically in the bottom quarter • Driving surfaces – blades – flat prior to 18th century, curved thereafter • Water – large volume, low head • Efficiency – about 20% prior to 18th century and later 50 to 60%	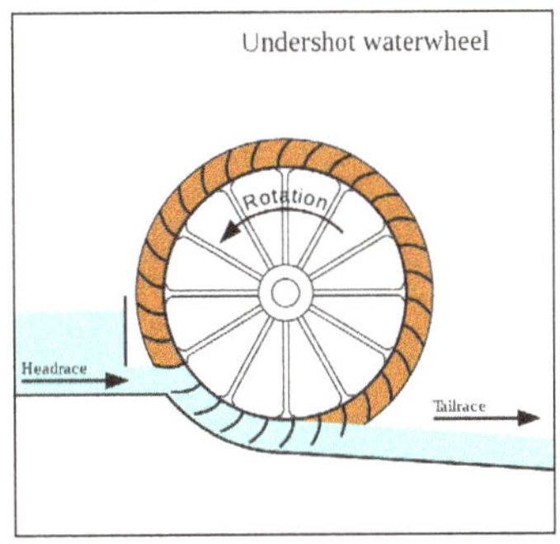 Diagram of undershot waterwheel showing headrace, tailrace, and water.
Breastshot • Vertical wheel with horizontal axle • The water hits the wheel roughly central, typically between one quarter and three quarters of the height. • Driving surfaces – buckets – carefully shaped to ensure that the water enters smoothly • Water – large volume, moderate head • Efficiency – 50 to 60%	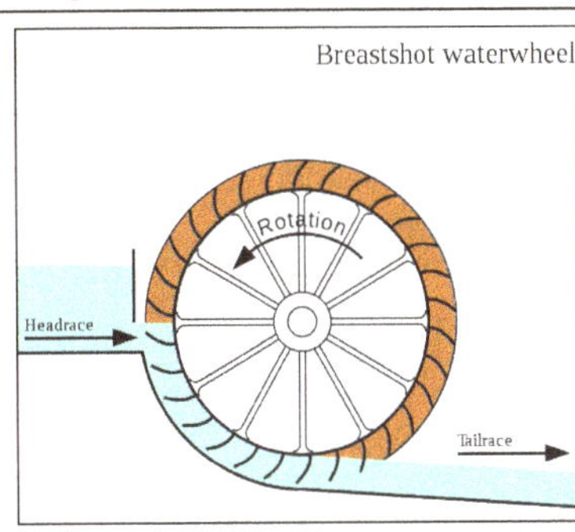 Diagram of breastshot waterwheel showing headrace, tailrace, and water.
Backshot (also known as pitchback) • Vertical wheel with horizontal axle • The water hits near the top of the wheel and before the axle so that it turns back towards the head race • Driving surfaces – buckets • Water – low volume, large head • Efficiency – 80 to 90%	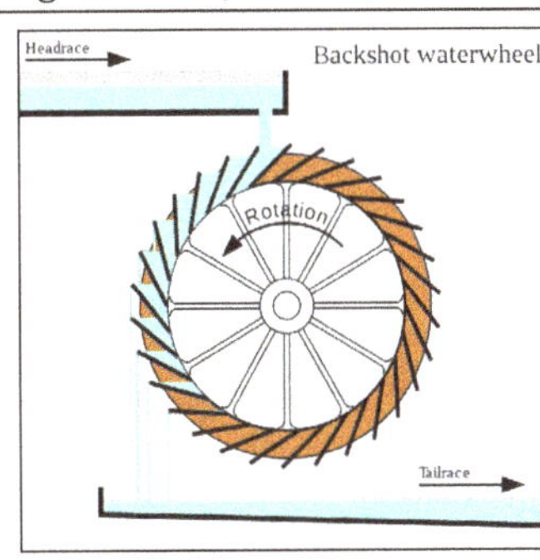 Diagram of backshot waterwheel showing headrace, tailrace, water, and spillage.

Overshot • Vertical wheel with horizontal axle • The water hits near the top of the wheel and in front of the axle so that it turns away from the head race • Driving surfaces – buckets • Water – low volume, large head • Efficiency – 80 to 90%	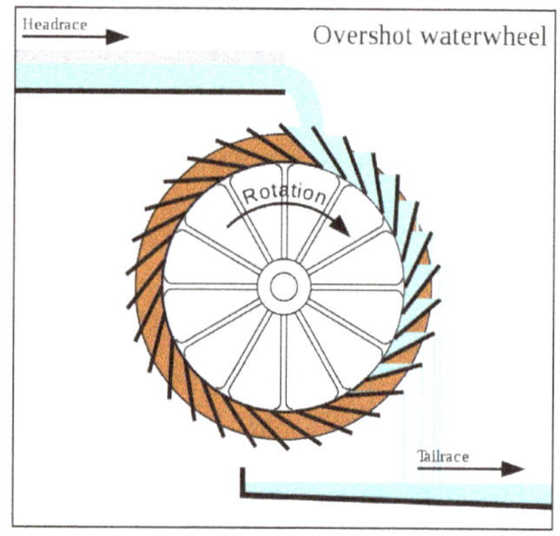 Diagram of overshot waterwheel showing headrace, tailrace, water, and spillage.

Overshot and backshot water wheels are typically used where the available height difference is more than a couple of meters. Breastshot wheels are more suited to large flows with a moderate head. Undershot and stream wheel use large flows at little or no head.

There is often an associated millpond, a reservoir for storing water and hence energy until it is needed. Larger heads store more potential energy for the same amount of water so the reservoirs for overshot and backshot wheels tend to be smaller than for breast shot wheels.

Overshot and pitchback water wheels are suitable where there is a small stream with a height difference of more than 2 metres (6.5 ft), often in association with a small reservoir. Breastshot and undershot wheels can be used on rivers or high volume flows with large reservoirs.

Vertical Axis

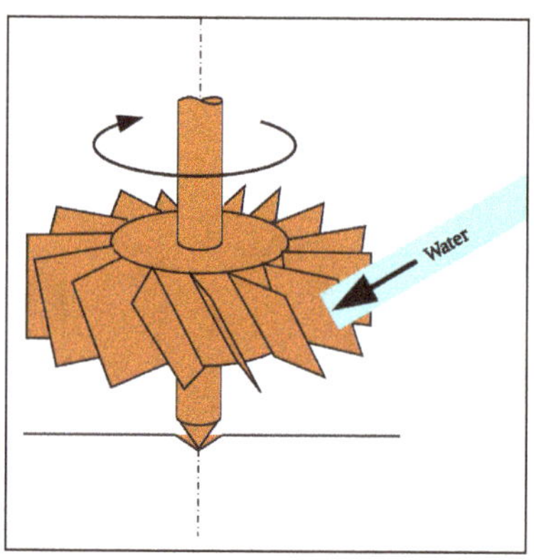

Diagram of vertical axis water mill.

A horizontal Wheel with a Vertical Axle. Commonly called a tub wheel, Norse mill or Greek mill, the horizontal wheel is a primitive and inefficient form of the modern turbine. However if it delivers the required power then the efficiency is of secondary importance. It is usually mounted inside a mill building below the working floor. A jet of water is directed on to the paddles of the water wheel, causing them to turn. This is a simple system usually without gearing so that the vertical axle of the water wheel becomes the drive spindle of the mill.

The earliest known reference to water wheels dates to about 400 BCE, and the earliest horizontal axis wheels date to about 200 BCE, so vertical axis mills pre-date horizontal axis mills by about two centuries.

Stream

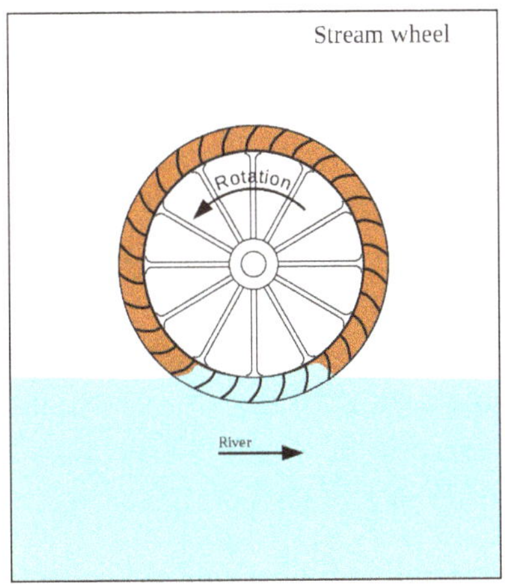

Diagram of stream shot waterwheel.

A stream wheel is a vertically mounted water wheel that is rotated by the water in a water course striking paddles or blades at the bottom of the wheel. This type of water wheel is the oldest type of horizontal axis wheel. They are also known as free surface wheels because the water is not constrained by millraces or wheel pit.

Stream wheels are cheaper and simpler to build, and have less of an environmental impact, than other type of wheel. They do not constitute a major change of the river. Their disadvantages are their low efficiency, which means that they generate less power and can only be used where the flow rate is sufficient. A typical flat board undershot wheel uses about 20 percent of the energy in the flow of water striking the wheel as measured by English civil engineer John Smeaton in the 18th century. More modern wheels have higher efficiencies. Stream wheels gain little or no advantage from head, a difference in water level.

Stream wheels mounted on floating platforms are often referred to as ship wheels and the mill as a ship mill. The earliest were probably constructed by the Byzantine general Belisarius during the siege of Rome in 537. Later they were sometimes mounted immediately downstream from bridges where the flow restriction of the bridge piers increased the speed of the current.

Historically they were very inefficient but major advances were made in the eighteenth century.

Undershot Wheel

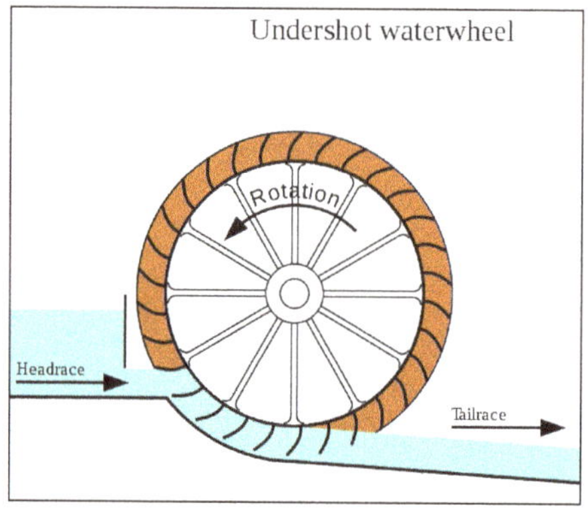

Diagram of undershot waterwheel showing headrace, tailrace, and water.

The sound of the Otley waterwheel, at Manchester Museum of Science and Industry.

An undershot wheel is a vertically mounted water wheel with a horizontal axle that is rotated by the water from a low weir striking the wheel in the bottom quarter. Most of the energy gain is from the movement of the water and comparatively little from the head. They are similar in operation and design to stream wheels.

The term undershot is sometimes used with related but different meanings:

- All wheels where the water passes under the wheel.
- Wheels where the water enters in the bottom quarter.
- Wheels where paddles are placed into the flow of a stream.

This is the oldest type of vertical water wheel.

Breastshot Wheel

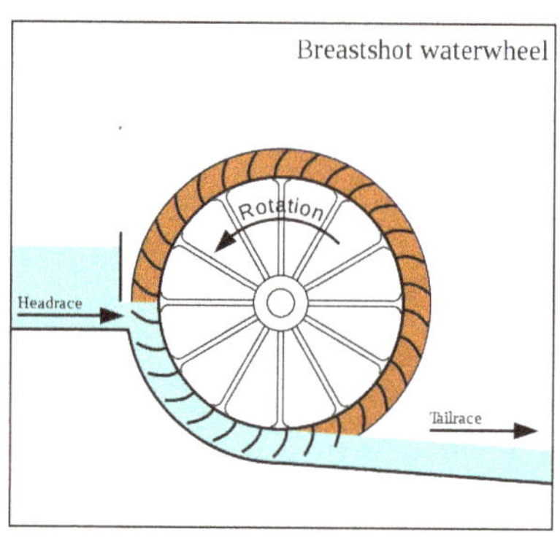

Diagram of breastshot waterwheel showing headrace, tailrace, and water.

The word breastshot is used in a variety of ways. Some authors restrict the term to wheels where the water enters at about the 10 o'clock position, others 9 o'clock, and others for a range of heights. In this part it is used for wheels where the water entry is significantly above the bottom and significantly below the top, typically the middle half.

They are characterised by:

- Buckets carefully shaped to minimise turbulence as water enters.
- Buckets ventilated with holes in the side to allow air to escape as the water enters.
- A masonry "apron" closely conforming to the wheel face, which helps contain the water in the buckets as they progress downwards.

Both kinetic (movement) and potential (height and weight) energy are utilised.

The small clearance between the wheel and the masonry requires that a breastshot wheel has a good trash rack ('screen' in British English) to prevent debris from jamming between the wheel and the apron and potentially causing serious damage.

Breastshot wheels are less efficient than overshot and backshot wheels but they can handle high flow rates and consequently high power. They are preferred for steady, high-volume flows such as are found on the Fall Line of the North American East Coast. Breastshot wheels are the most common type in the United States of America and are said to have powered the industrial revolution.

Backshot Wheel

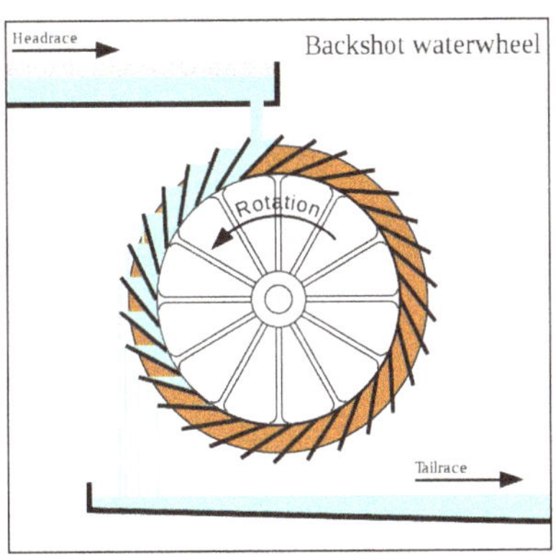

Diagram of backshot waterwheel showing headrace, tailrace, water, and spillage.

A backshot wheel (also called pitchback) is a variety of overshot wheel where the water is introduced just before the summit of the wheel. In many situations it has the advantage that the bottom of the wheel is moving in the same direction as the water in the tail race which makes it more efficient. It also performs better than an overshot wheel in flood conditions when the water level may submerge the bottom of the wheel. It will continue to rotate until the water in the wheel pit rises quite high on the wheel. This makes the technique particularly suitable for streams that experience significant variations in flow and reduces the size, complexity and hence cost of the tail race.

Backshot wheel at New Lanark World Heritage Site, Scotland.

The direction of rotation of a backshot wheel is the same as that of a breastshot wheel but in other respects it is very similar to the overshot wheel.

Overshot Wheel

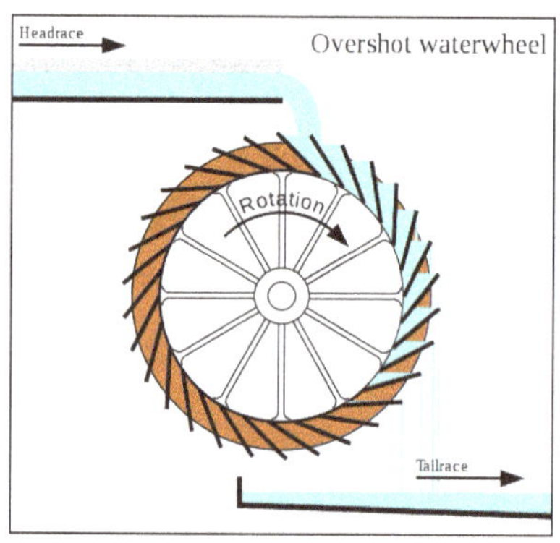

Diagram of overshot waterwheel showing headrace, tailrace, water, and spillage.

A vertically mounted water wheel that is rotated by water entering buckets just past the top of the wheel is said to be overshot. The term is sometimes, erroneously, applied to backshot wheels where the water goes down behind the wheel.

A typical overshot wheel has the water channelled to the wheel at the top and slightly beyond the axle. The water collects in the buckets on that side of the wheel, making it heavier than the other "empty" side. The weight turns the wheel, and the water flows out into the tail-water when the wheel rotates enough to invert the buckets. The overshot design is very efficient, it can achieve 90%, and does not require rapid flow.

Nearly all of the energy is gained from the weight of water lowered to the tail race although a small contribution may be made by the kinetic energy of the water entering the wheel. They are suited to larger heads than the other type of wheel so they are ideally suited to hilly country. However even the largest water wheel, the Laxey Wheel in the Isle of Man, only utilises a head of around 30 m (100 ft). The world's largest head turbines, Bieudron Hydroelectric Power Station in Switzerland, utilise about 1,869 m (6,132 ft).

Overshot wheels require a large head compared to other types of wheel which usually means significant investment in constructing the head race. Sometimes the final approach of the water to the wheel is along a flume or penstock, which can be lengthy.

Hybrid

Overshot and Backshot

One of Finch Foundry's water wheels.

Some wheels are overshot at the top and backshot at the bottom thereby potentially combining the best features of both types. The photograph shows an example at Finch Foundry in Devon, UK. The head race is the overhead timber structure and a branch to the left supplies water to the wheel. The water exits from under the wheel back into the stream.

Reversible

The Anderson Mill of Texas is undershot, backshot, and overshot using two sources of water. This allows the direction of the wheel to be reversed.

A special type of overshot/backshot wheel is the reversible water wheel. This has two sets of blades or buckets running in opposite directions, so that it can turn in either direction depending on which side the water is directed. Reversible wheels were used in the mining industry in order to power various means of ore conveyance. By changing the direction of the wheel, barrels or baskets of ore could be lifted up or lowered down a shaft or inclined plane. There was usually a cable drum or a chain basket (German: Kettenkorb) on the axle of the wheel. It is essential that the wheel have braking equipment to be able to stop the wheel (known as a braking wheel). The oldest known drawing of a reversible water wheel was by Georgius Agricola and dates to 1556.

Suspension Wheels and Rim-gears

The suspension wheel with rim-gearing at the Portland Basin Canal Warehouse.

Two early improvements were suspension wheels and rim gearing. Suspension wheels are constructed in the same manner as a bicycle wheel, the rim being supported under tension from the hub- this led to larger lighter wheels than the former design where the heavy spokes were under compression. Rim-gearing entailed adding a notched wheel to the rim or shroud of the wheel. A stub gear engaged the rim-gear and took the power into the mill using an independent line shaft. This removed the rotative stress from the axle which could thus be lighter, and also allowed more flexibility in the location of the power train. The shaft rotation was geared up from that of the wheel which led to less power loss. An example of this design pioneered by Thomas Hewes and refined by William Fairburn can be seen at the 1849 restored wheel at the Portland Basin Canal Warehouse.

Efficiency

Overshot (and particularly backshot) wheels are the most efficient type; a backshot steel wheel can be more efficient (about 60%) than all but the most advanced and well-constructed turbines. In some situations an overshot wheel is preferable to a turbine.

The development of the hydraulic turbine wheels with their improved efficiency (>67%) opened up an alternative path for the installation of water wheels in existing mills, or redevelopment of abandoned mills.

The Power of a Wheel

The energy available to the wheel has two components:

- Kinetic energy – depends on how fast the water is moving when it enters the wheel.
- Potential energy – depends on the change in height of the water between entry to and exit from the wheel.

The kinetic energy can be accounted for by converting it into an equivalent head, the velocity head, and adding it to the actual head. For still water the velocity head is zero, and to a good approximation it is negligible for slowly moving water, and can be ignored. The velocity in the tail race is not taken into account because for a perfect wheel the water would leave with zero energy which requires zero velocity. That is impossible, the water has to move away from the wheel, and represents an unavoidable cause of inefficiency.

The power is how fast that energy is delivered which is determined by the flow rate.

Quantities and units:

- η efficiency.
- ρ density of water (1000 kg/m^3).
- A cross sectional area of the channel (m^2).
- D diameter of wheel (m).
- P power (W).
- d distance (m).
- g strength of gravity (9.81 m/s^2 = 9.81 N/kg).
- h head (m).
- h_p pressure head, the difference in water levels (m).
- h_v velocity head (m).
- k velocity correction factor. 0.9 for smooth channels.
- v velocity (m/s).
- $\dot{q}$ volume flow rate (m^3/s).
- t time (s).

Measurements

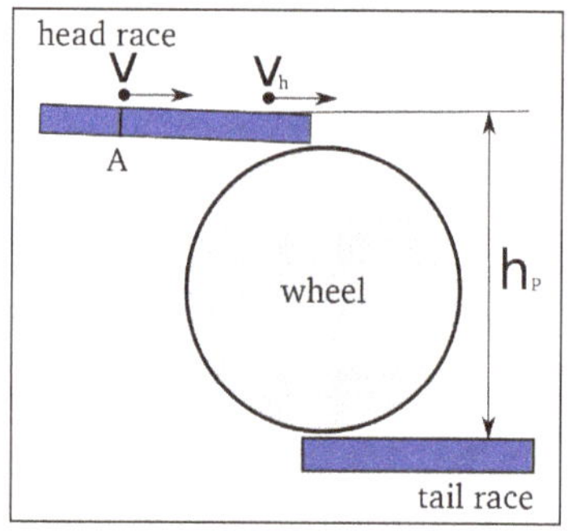

How to measure the head and flow rate of a water wheel.

Pressure head h_p = is the difference in height between the head race and tail race water surfaces.

Velocity head h_v = is calculated from the velocity of the water in the head race at the same place as the pressure head is measured from.

The velocity (speed) v can be measured by the pooh sticks method, timing a floating object over a measured distance. The water at the surface moves faster than water nearer to the bottom and sides so a correction factor should be applied as in the formula below.

There are many ways to measure the volume flow rate. Two of the simplest are:

- From the cross sectional area and the velocity. They must be measured at the same place but that can be anywhere in the head or tail races. It must have the same amount of water going through it as the wheel.
- It is sometimes practicable to measure the volume flow rate by the bucket and stop watch method.

Formulae

Quantity	Formula
Power	$P = \eta \cdot \rho \cdot g \cdot h \cdot \dot{q}$
Effective head	$h = h_p + h_v$
Velocity head	$h_v = \dfrac{v^2}{2 \cdot g}$
Volume flow rate	$\dot{q} = A \cdot v$
Water velocity (speed)	$v = k \cdot \dfrac{d}{t}$

Rules of Thumb

Breast and Overshot

Quantity	Approximate Formula
Power (assuming 70% efficiency)	$P = 7000 \cdot \dot{q} \cdot h$
Optimal rotational speed	$\frac{21}{\sqrt{D}}$ rpm

Traditional Undershot Wheels

Quantity	Approximate Formula
Power (assuming 20% efficiency)	$P = 100 \cdot A \cdot v^3$
Optimal rotational speed	$\frac{9 \cdot v}{D}$ rpm

Hydraulic Wheel Part Reaction Turbine

A parallel development is the hydraulic wheel/part reaction turbine that also incorporates a weir into the centre of the wheel but uses blades angled to the water flow. The WICON-Stem Pressure Machine (SPM) exploits this flow. Estimated efficiency 67%.

The University of Southampton School of Civil Engineering and the Environment in the UK has investigated both types of Hydraulic wheel machines and has estimated their hydraulic efficiency and suggested improvements, i.e. The Rotary Hydraulic Pressure Machine. (Estimated maximum efficiency 85%).

These type of water wheels have high efficiency at part loads / variable flows and can operate at very low heads, < 1 m (3 ft 3 in). Combined with direct drive Axial Flux Permanent Magnet Alternators and power electronics they offer a viable alternative for low head hydroelectric power generation.

Water-lifting

In water-raising devices rotary motion is typically more efficient than machines based on oscillating motion.

The compartmented water wheel comes in two basic forms, the wheel with compartmented body (Latin *tympanum*) and the wheel with compartmented rim or a rim with separate, attached containers. The wheels could be either turned by the flow of water, men treading on its outside or by animals by means of a sakia gear. While the tympanum had a large discharge capacity, it could lift the water only to less than the height of its own radius and required a large torque for rotating. These constructional deficiencies were overcome by the wheel with a compartmented rim which was a less heavy design with a higher lift.

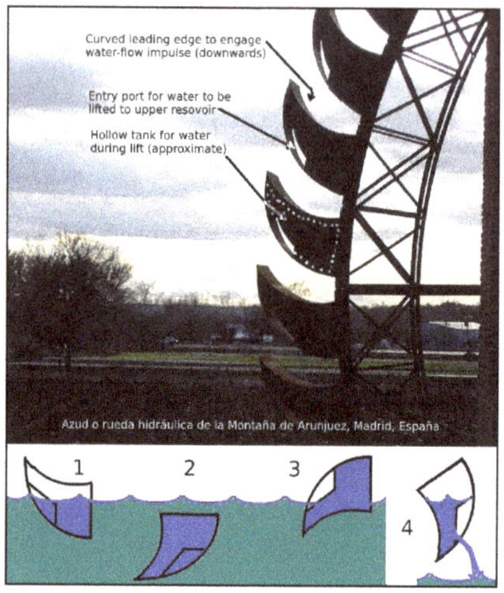

Detail of *azud* at Aranjuez, Spain.

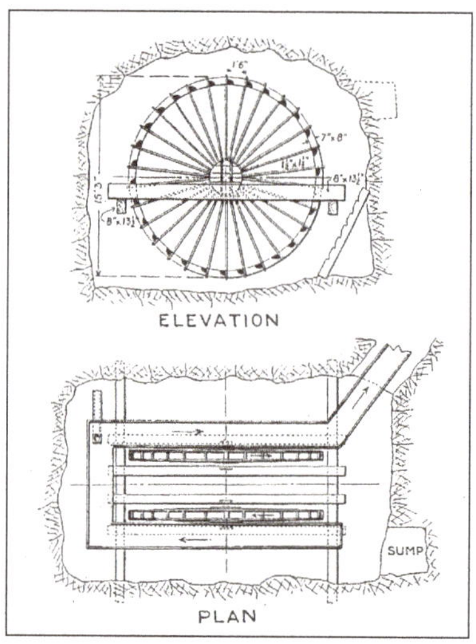

Drainage wheel from Rio Tinto mines.

Ptolemaic Egypt

The earliest literary reference to a water-driven, compartmented wheel appears in the technical treatise *Pneumatica* of the Greek engineer Philo of Byzantium. In his *Parasceuastica*, Philo advises the use of such wheels for submerging siege mines as a defensive measure against enemy sapping. Compartmented wheels appear to have been the means of choice for draining dry docks in Alexandria under the reign of Ptolemy IV. Several Greek papyri of the 3rd to 2nd century BCE mention the use of these wheels, but don't give further details. The non-existence of the device in the Ancient Near East before Alexander's conquest can be deduced from its pronounced absence from the otherwise rich oriental iconography on irrigation practices. Unlike other water-lifting devices and pumps of the period though, the invention of the compartmented wheel cannot be traced to any particular Hellenistic engineer and may have been made in the late 4th century BCE in a rural context away from the metropolis of Alexandria.

The earliest depiction of a compartmented wheel is from a tomb painting in Ptolemaic Egypt which dates to the 2nd century BCE. It shows a pair of yoked oxen driving the wheel via a sakia gear, which is here for the first time attested, too. The Greek sakia gear system is already shown fully developed to the point that "modern Egyptian devices are virtually identical". It is assumed that the scientists of the Museum of Alexandria, at the time the most active Greek research center, may have been involved in its invention. An episode from the Alexandrian War in 48 BCE tells of how Caesar's enemies employed geared water wheels to pour sea water from elevated places on the position of the trapped Romans.

Around 300 CE, the noria was finally introduced when the wooden compartments were replaced with inexpensive ceramic pots that were tied to the outside of an open-framed wheel.

References

- Water-turbine-design, hydro-energy: alternative-energy-tutorials.com, Retrieved 11 August, 2019
- Petkewich, rachel (2004). "technology solutions: creating electricity with undammed hydropower". Environmental science & technology. 38 (3): 55a–56a. Doi:10.1021/es0403716. Pmid 14968846
- Francis-turbines-components-application, practical-guide: theconstructor.org, Retrieved 11 January, 2019
- Darling, david. "turgo turbine". Encyclopedia of alternative energy. The worlds of dvid darling. Retrieved 7 march 2016
- Pelton-turbine, encyclopedia: energyeducation.ca, Retrieved 12 July, 2019
- Thomson, ross (2009). Structures of change in the mechanical age: technological invention in the united states 1790-1865. Baltimore, md: the johns hopkins university press. P. 34. Isbn 978-0-8018-9141-0

5

Hydropower Plants

A few of the common types of hydropower plants are conventional hydroelectric power stations, pumped-storage hydroelectric power stations, run-of-the-river hydroelectric power stations and tidal power stations. The topics elaborated in this chapter will help in gaining a better perspective about these types of hydropower plants.

CONVENTIONAL HYDROELECTRIC POWER STATIONS

Hydropower plants harness water's energy and use simple mechanics to convert that energy into electricity. Hydropower plants are actually based on a rather simple concept - water flowing through a dam turns a turbine, which turns a generator.

Here are the basic components of a conventional hydropower plant:

The shaft that connects the turbine and generator.

- Dam: Most hydropower plants rely on a dam that holds back water, creating a large reservoir. Often, this reservoir is used as a recreational lake, such as Lake Roosevelt at the Grand Coulee Dam in Washington State.

- Intake: Gates on the dam open and gravity pulls the water through the penstock, a pipeline that leads to the turbine. Water builds up pressure as it flows through this pipe.
- Turbine: The water strikes and turns the large blades of a turbine, which is attached to a generator above it by way of a shaft. The most common type of turbine for hydropower plants is the Francis Turbine, which looks like a big disc with curved blades. A turbine can weigh as much as 172 tons and turn at a rate of 90 revolutions per minute (rpm), according to the Foundation for Water & Energy Education (FWEE).
- Generators: As the turbine blades turn, so do a series of magnets inside the generator. Giant magnets rotate past copper coils, producing alternating current (AC) by moving electrons.
- Transformer: The transformer inside the powerhouse takes the AC and converts it to higher-voltage current.
- Power lines: Out of every power plant come four wires: the three phases of power being produced simultaneously plus a neutral or ground common to all three.
- Outflow: Used water is carried through pipelines, called tailraces, and re-enters the river downstream.

The water in the reservoir is considered stored energy. When the gates open, the water flowing through the penstock becomes kinetic energy because it's in motion. The amount of electricity that is generated is determined by several factors. Two of those factors are the volume of water flow and the amount of hydraulic head. The head refers to the distance between the water surface and the turbines. As the head and flow increase, so does the electricity generated. The head is usually dependent upon the amount of water in the reservoir.

Three Gorges Dam

The Three Gorges Dam is a hydroelectric gravity dam that spans the Yangtze River by the town of Sandouping, in Yiling District, Yichang, Hubei province, China. The Three Gorges Dam has been the world's largest power station in terms of installed capacity (22,500 MW) since 2012. In 2014, the dam generated 98.8 terawatt-hours (TWh) and had the world record, but was surpassed by the Itaipú Dam, which set the new world record in 2016, producing 103.1 TWh.

Except for the locks, the dam project was completed and fully functional as of July 4, 2012, when the last of the main water turbines in the underground plant began production. The ship lift was complete in December 2015. Each main water turbine has a capacity of 700 MW. The dam body was completed in 2006. Coupling the dam's 32 main turbines with two smaller generators (50 MW each) to power the plant itself, the total electric generating capacity of the dam is 22,500 MW.

As well as producing electricity, the dam is intended to increase the Yangtze River's shipping capacity and reduce the potential for floods downstream by providing flood storage space. China regards the project as monumental as well as a success socially and economically, with the design of state-of-the-art large turbines, and a move toward limiting greenhouse gas emissions. However, the dam flooded archaeological and cultural sites, displaced some 1.3 million people, and had caused significant ecological changes including an increased risk of landslides. The dam has been controversial both domestically and abroad.

Composition and Dimensions

Model of the Three Gorges Dam showing the ship lift and the ship lock. The ship lift is to the right of the dam body with its own designated waterway. The ship locks are to the right (northeast) of the ship lift.

Earthfill south dam in foreground with view along main dam. The wall beyond is to separate spillway and turbine flows from the lock and ship lift upstream approach channel. A similar separation is used on the downstream side, seen partially in the preceding image.

Made of concrete and steel, the dam is 2,335 m (7,661 ft) long and the top of the dam is 185 m (607 ft) above sea level. The project used 27.2×10^6 m^3 (35.6×10^6 cu yd) of concrete (mainly for the dam wall), used 463,000 T of steel (enough to build 63 Eiffel Towers), and moved about 102.6×10^6 m^3 (134.2×10^6 cu yd) of earth. The concrete dam wall is 181 m (594 ft) high above the rock basis.

When the water level is at its maximum of 175 m (574 ft) above sea level, 110 m (361 ft) higher than the river level downstream, the dam reservoir is on average about 660 km (410 mi) in length and 1.12 km (3,675 ft) in width. It contains 39.3 km^3 (31,900,000 acre·ft) of water and has a total surface area of 1,045 km² (403 sq mi). On completion, the reservoir flooded a total area of 632 km² (244 sq mi) of land, compared to the 1,350 km² (520 sq mi) of reservoir created by the Itaipu Dam.

Power Generation and Distribution

Generating Capacity

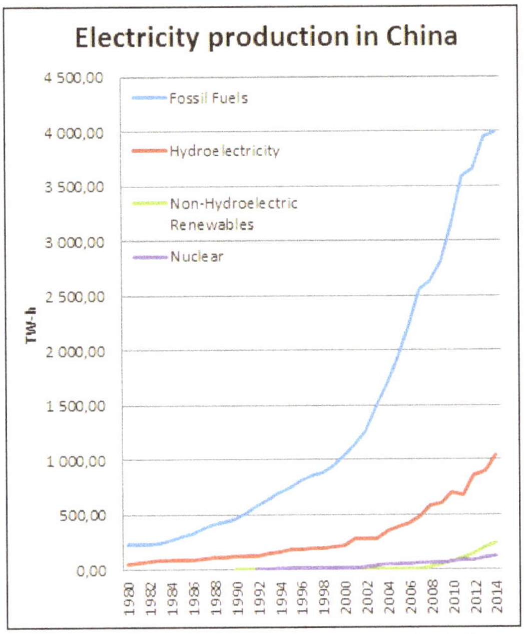

Electricity production in China by source. Compare: The fully completed Three Gorges dam will contribute about 100 TWh of generation per year.

Power generation is managed by China Yangtze Power, a listed subsidiary of China Three Gorges Corporation (CTGC)—a Central Enterprise SOE administered by SASAC. The Three Gorges Dam is the world's largest capacity hydroelectric power station with 34 generators: 32 main generators, each with a capacity of 700 MW, and two plant power generators, each with capacity of 50 MW, making a total capacity of 22,500 MW. Among those 32 main generators, 14 are installed in the north side of the dam, 12 in the south side, and the remaining six in the underground power plant in the mountain south of the dam. Annual electricity generation in 2015 was 87 TWh, which is 20 times more than the Hoover Dam.

Generators

The main generators weigh about 6,000 tonnes each and are designed to produce more than 700 MW of power. The designed head of the generator is 80.6 meters (264 ft). The flow rate varies between 600–950 cubic metres per second (21,000–34,000 cu ft/s) depending on the head available. The greater the head, the less water needed to reach full power. Three Gorges uses Francis turbines. Turbine diameter is 9.7/10.4 m

(VGS design/Alstom's design) and rotation speed is 75 revolutions per minute. This means that in order to generate power at 50 Hz, the generator rotors have 80 poles. Rated power is 778 MVA, with a maximum of 840 MVA and a power factor of 0.9. The generator produces electrical power at 20 kV. The electricity generated is then stepped-up to 500 kV for transmission at 50 Hz. The outer diameter of the generator stator is 21.4/20.9 m. The inner diameter is 18.5/18.8 m. The stator, the biggest of its kind, is 3.1/3 m in height. Bearing load is 5050/5500 tonnes. Average efficiency is over 94%, and reaches 96.5%.

Three Gorges Dam Francis turbine.

The generators were manufactured by two joint ventures: one of them Alstom, ABB Group, Kvaerner, and the Chinese company Harbin Motor; the other Voith, General Electric, Siemens (abbreviated as VGS), and the Chinese company Oriental Motor. The technology transfer agreement was signed together with the contract. Most of the generators are water-cooled. Some newer ones are air-cooled, which are simpler in design and manufacture and are easier to maintain.

Generator Installation Progress

The first north side main generator (No. 2) started on July 10, 2003; the north side became completely operational September 7, 2005, with the implementation of generator No. 9. Full power (9,800 MW) was only reached on October 18, 2006, after the water level reached 156 m.

The 12 south side main generators are also in operation. No. 22 began operation on June 11, 2007, and No. 15 started up on October 30, 2008. The sixth (No. 17) began operation on December 18, 2007, raising capacity to 14.1 GW, finally surpassing Itaipu (14.0 GW), to become the world's largest hydro power plant by capacity.

As of May 23, 2012, when the last main generator, No. 27, finished its final test, the six underground main generators are also in operation, raising capacity to 22.5 GW. After nine years of construction, installation and testing, the power plant is now fully operational.

Output Milestones

Annual Production of Electricity		
Year	**Number of installed units**	**TWh**
2003	6	8.607
2004	11	39.155
2005	14	49.090
2006	14	49.250
2007	21	61.600
2008	26	80.812
2009	26	79.470
2010	26	84.370
2011	29	78.290
2012	32	98.100
2013	32	83.270
2014	32	98.800
2015	32	87.000
2016	32	93.500
2017	32	97.600
2018	32	100+

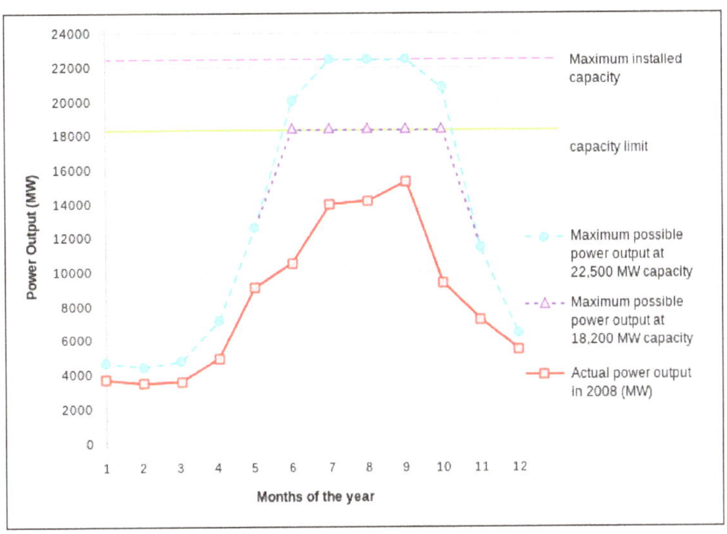

Three Gorges Dam annual power output.

By August 16, 2011, the plant had generated 500 TWh of electricity. In July 2008 it generated 10.3 TWh of electricity, its first month over 10 TWh. On June 30, 2009, after the river flow rate increased to over 24,000 m^3, all 28 generators were switched on, producing only 16,100 MW because the head available during flood season is insufficient. During an August 2009 flood, the plant first reached its maximum output for a short period.

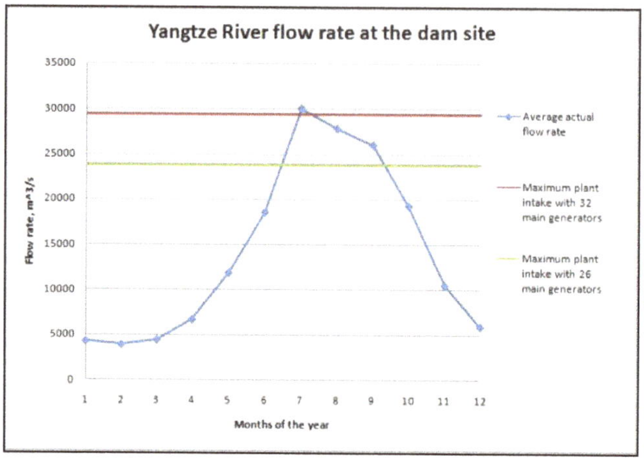

Yangtze River flow rate comparing to the dam intake capacity.

During the November to May dry season, power output is limited by the river's flow rate, as seen in the diagrams on the right. When there is enough flow, power output is limited by plant generating capacity. The maximum power-output curves were calculated based on the average flow rate at the dam site, assuming the water level is 175 m and the plant gross efficiency is 90.15%. The actual power output in 2008 was obtained based on the monthly electricity sent to the grid.

The Three Gorges Dam reached its design-maximum reservoir water level of 175 m (574 ft) for the first time on October 26, 2010, in which the intended annual power-generation capacity of 84.7 TWh was realized. In 2012, the dam's 32 generating units generated a record 98.1 TWh of electricity, which accounts for 14% of China's total hydro generation.

Distribution

The State Grid Corporation and China Southern Power Grid paid a flat rate of ¥250 per MWh (US$35.7) until July 2, 2008. Since then, the price has varied by province, from ¥228.7–401.8 per MWh. Higher-paying customers, such as Shanghai, receive priority. Nine provinces and two cities consume power from the dam.

Power distribution and transmission infrastructure cost about 34.387 billion Yuan. Construction was completed in December 2007, one year ahead of schedule.

Power is distributed over multiple 500 kilovolt (kV) transmission lines. Three direct current (DC) lines to the East China Grid carry 7,200 MW: Three Gorges – Shanghai (3,000 MW), HVDC Three Gorges – Changzhou (3,000 MW), and HVDC Gezhouba – Shanghai (1,200 MW). The alternating current (AC) lines to the Central China Grid have a total capacity of 12,000 MW. The DC transmission line HVDC Three Gorges – Guangdong to the South China Grid has a capacity of 3,000 MW.

The dam was expected to provide 10% of China's power. However, electricity demand has increased more quickly than previously projected. Even fully operational, on average, it supports only about 1.7% of electricity demand in China in the year of 2011, when the Chinese electricity demand reached 4692.8 TWh.

Environmental Impact

Emissions

According to the National Development and Reform Commission of China, 366 grams of coal would produce 1 kWh of electricity during 2006. At full power, Three Gorges reduces coal consumption by 31 million tonnes per year, avoiding 100 million tonnes of greenhouse gas emissions, millions of tonnes of dust, one million tonnes of sulfur dioxide, 370,000 tonnes of nitric oxide, 10,000 tonnes of carbon monoxide, and a significant amount of mercury. Hydropower saves the energy needed to mine, wash, and transport the coal from northern China.

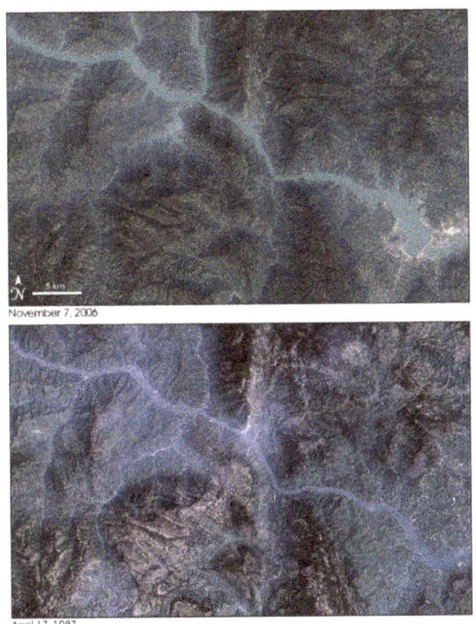

Satellite map showing areas flooded by the Three Gorges reservoir. Compare November 7, 2006 (above) with April 17, 1987 (below).

Flood mark on Yangtze river.

From 2003 to 2007, power production equaled that of 84 million tonnes of standard coal, reducing carbon dioxide by 190 million tonnes, sulfur dioxide by 2.29 million tonnes, and nitrogen oxides by 980,000 tonnes.

The dam increased the Yangtze's barge capacity sixfold, reducing carbon dioxide emission by 630,000 tonnes. From 2004 to 2007, a total of 198 million tonnes of goods passed through the ship locks. Compared to using trucking, barges reduced carbon dioxide emission by ten million tonnes and lowered costs by 25%.

Erosion and Sedimentation

Two hazards are uniquely identified with the dam. One is that sedimentation projections are not agreed upon, and the other is that the dam sits on a seismic fault. At current levels, 80% of the land in the area is experiencing erosion, depositing about 40 million tons of sediment into the Yangtze annually. Because the flow is slower above the dam, much of this sediment will now settle there instead of flowing downstream, and there will be less sediment downstream.

The absence of silt downstream has three effects:

- Some hydrologists expect downstream riverbanks to become more vulnerable to flooding.
- Shanghai, more than 1,600 km (990 mi) away, rests on a massive sedimentary plain. The "arriving silt—so long as it does arrive—strengthens the bed on which Shanghai is built the less the tonnage of arriving sediment the more vulnerable is this biggest of Chinese cities to inundation."
- Benthic sediment buildup causes biological damage and reduces aquatic biodiversity.

Landslides

Erosion in the reservoir, induced by rising water, causes frequent major landslides that have led to noticeable disturbance in the reservoir surface, including two incidents in May 2009 when somewhere between 20,000 and 50,000 cubic metres (26,000 and 65,000 cu yd) of material plunged into the flooded Wuxia Gorge of the Wu River. Also, in the first four months of 2010, there were 97 significant landslides.

Waste Management

The dam catalyzed improved upstream wastewater treatment around Chongqing and its suburban areas. According to the Ministry of Environmental Protection, as of April 2007, more than 50 new plants could treat 1.84 million tonnes per day, 65% of the total need. About 32 landfills were added, which could handle 7,664.5 tonnes of solid waste every day. Over one billion tons of wastewater are released annually into the river, which was more likely to be swept away before the reservoir was created. This has left the water looking stagnant, polluted and murky.

Zigui County seat source water protection area in Maoping Town, a few kilometers upstream of the dam.

Collecting garbage at the Dam's southeast corner.

Forest Cover

In 1997, the Three Gorges area had 10% forestation, down from 20% in the 1950s.

Research by the United Nations Food and Agriculture Organization suggested that the Asia-Pacific region would, overall, gain about 6,000 km^2 (2,300 sq mi) of forest by 2008. That is a significant change from the 13,000 km^2 (5,000 sq mi) net loss of forest each year in the 1990s. This is largely due to China's large reforestation effort. This accelerated after the 1998 Yangtze River floods convinced the government that it must restore tree cover, especially in the Yangtze's basin upstream of the Three Gorges Dam.

Wildlife

Concerns about the potential wildlife impact of the dam predate the National People's Congress's approval in 1992. This region has long been known for its rich biodiversity. It is home to 6,388 species of plants, which belong to 238 families and 1508 genera. Of these plant species, 57 percent are endangered. These rare species are also used as ingredients in traditional Chinese medicines. Already, the percentage of forested area in the region surrounding the Three Gorges Dam has dropped from twenty percent in

1950 to less than ten percent as of 2002, negatively affecting all plant species in this locality. The region also provides habitats to hundreds of freshwater and terrestrial animal species. Freshwater fish are especially affected by dams due to changes in the water temperature and flow regime. Many other fish are hurt in the turbine blades of the hydroelectric plants as well. This is particularly detrimental to the ecosystem of the region because the Yangtze River basin is home to 361 different fish species and accounts for twenty-seven percent of all endangered freshwater fish species in China. Other aquatic species have been endangered by the dam, particularly the baiji, or Chinese river dolphin, now extinct. In fact, Government Chinese scholars even claim that the Three Gorges Dam directly caused the extinction of the baiji.

Of the 3,000 to 4,000 remaining critically endangered Siberian crane, a large number currently spend the winter in wetlands that will be destroyed by the Three Gorges Dam. The dam contributed to the functional extinction of the baiji Yangtze river dolphin. Though it was close to this level even at the start of construction, the dam further decreased its habitat and increased ship travel, which are among the factors causing what will be its ultimate demise. In addition, populations of the Yangtze sturgeon are guaranteed to be "negatively affected" by the dam.

Terrestrial Impact

In 2010, NASA scientists calculated that shift of water mass stored by the dams would increase the length of the Earth's day by 0.06 microseconds and make the Earth slightly more round in the middle and flat on the poles.

Floods, Agriculture and Industry

An important function of the dam is to control flooding, which is a major problem for the seasonal river of the Yangtze. Millions of people live downstream of the dam, with many large, important cities like Wuhan, Nanjing, and Shanghai situated adjacent to the river. Plenty of farm land and China's most important industrial area are built beside the river.

The reservoir's flood storage capacity is 22 cubic kilometres (5.3 cu mi; 18,000,000 acre·ft). This capacity will reduce the frequency of major downstream flooding from once every 10 years to once every 100 years. The dam is expected to minimize the effect of even a "super" flood. In 1954, the river flooded 193,000 km^2 (74,500 sq mi), killing 33,169 people and forcing 18,884,000 people to move. The flood covered Wuhan, a city of eight million people, for over three months, and the Jingguang Railway was out of service for more than 100 days. The 1954 flood carried 50 cubic kilometres (12 cu mi) of water. The dam could only divert the water above Chenglingji, leaving 30 to 40 km^3 (7.2 to 9.6 cu mi) to be diverted. Also, the dam cannot protect against some of the large tributaries downstream, including the Xiang, Zishui, Yuanshui, Lishui, Hanshui, and the Gan.

In 1998, a flood in the same area caused billions of dollars in damage; 2,039 km^2 (787 sq mi) of farm land were flooded. The flood affected more than 2.3 million people, killing 1,526. In early August 2009, the largest flood in five years passed through the dam site. The dam limited the water flow to less than 40,000 cubic metres (1,400,000 cu ft) per second, raising the upstream water level from 145.13 m (476.1 ft) on August 1, 2009, to 152.88 m (501.6 ft) on August 8, 2009. A full 4.27 km^3 (1.02 cu mi) of flood water was captured and the river flow was cut by as much as 15,000 m^3 (530,000 cu ft) per second.

The dam discharges its reservoir during the dry season between December and March every year. This increases the flow rate of the river downstream, and provides fresh water for agricultural and industrial usage. It also improves shipping conditions. The water level upstream drops from 175 to 145 m (574 to 476 ft), preparing for the rainy season. The water also powers the Gezhouba Dam downstream.

Since the filling of the reservoir in 2003, the Three Gorges Dam has supplied an extra 11 km^3 (2.6 cu mi) of fresh water to downstream cities and farms during the dry season.

During the 2010 South China floods in July, inflows at the Three Gorges Dam reached a peak of 70,000 m^3/s (2,500,000 cu ft/s), exceeding the peak during the 1998 Yangtze River Floods. The dam's reservoir rose nearly 3 m (9.8 ft) in 24 hours and reduced the outflow to 40,000 m^3/s (1,400,000 cu ft/s) in discharges downstream, effectively alleviating serious impacts on the middle and lower river.

Navigating the Dam

Locks

The installation of ship locks is intended to increase river shipping from ten million to 100 million tonnes annually, as a result transportation costs will be cut between 30 and 37%. Shipping will become safer, since the gorges are notoriously dangerous to navigate. Ships with much deeper draft will be able to navigate 2,400 kilometres (1,500 mi) upstream from Shanghai all the way to Chongqing. It is expected that shipping to Chongqing will increase fivefold.

Ship locks for river traffic to bypass the Three Gorges Dam, May 2004.

The other end of Three gorges dam lock, note the Bridge in the background.

Therearetwoseriesofshiplocksinstallednearthedam(30°50′12″N111°1′10″E30.83667°N 111.01944°E). Each of them is made up of five stages, with transit time at around four hours. Maximum vessel size is 10,000 tons. The locks are 280 m long, 35 m wide, and 5 m deep (918 × 114 × 16.4 ft). That is 30 m (98 ft) longer than those on the St Lawrence Seaway, but half as deep. Before the dam was constructed, the maximum freight capacity at the Three Gorges site was 18.0 million tonnes per year. From 2004 to 2007, a total of 198 million tonnes of freight passed through the locks. The freight capacity of the river increased six times and the cost of shipping was reduced by 25%. The total capacity of the ship locks is expected to reach 100 million tonnes per year.

These locks are staircase locks, whereby inner lock gate pairs serve as both the upper gate and lower gate. The gates are the vulnerable hinged type, which, if damaged, could temporarily render the entire flight unusable. As there are separate sets of locks for upstream and downstream traffic, this system is more water efficient than bi-directional staircase locks.

Ship Lift

The *shiplift*, a kind of elevator, can lift vessels of up to 3,000 tonnes, at a fraction of the time to transit the staircase locks.

In addition to the canal locks, there is a ship lift, a kind of elevator for vessels. The ship lift can lift ships of up to 3,000 tons. The vertical distance traveled is 113 m (371 ft), and the size of the ship lift's basin is 120 m × 18 m × 3.5 m (394 ft × 59 ft × 11 ft). The ship lift takes 30 to 40 minutes to transit, as opposed to the three to four hours for stepping through the locks. One complicating factor is that the water level can vary dramatically. The ship lift must work even if water levels vary by 12 meters (39 ft) on the lower side, and 30 m (98 ft) on the upper side.

The ship lift's design uses a helical gear system, to climb or descend a toothed rack. The ship lift was not yet complete when the rest of the project was officially opened on May 20, 2006. In November 2007, it was reported in the local media that construction of the ship lift started in October 2007.

In February 2012, *Xinhua* reported that the four towers that are to support the ship lift had almost been completed. The report said the towers had reached 189 m (620 ft) of the anticipated 195 m (640 ft), the towers would be completed by June 2012 and the entire shiplift in 2015.

As of May 2014, the ship lift was expected to be completed by July 2015. It was tested in December 2015 and announced complete in January 2016. Lahmeyer, the German firm that designed the ship lift, said it will take a vessel less than an hour to transit the lift. An article in Steel Construction says the actual time of the lift will be 21 minutes. It says that the expected dimensions of the 3,000 tonnes (3,000,000 kg) passenger vessels the ship lift's basin was designed to carry will be 84.5 by 17.2 by 2.65 metres (277.2 ft × 56.4 ft × 8.7 ft). The moving mass (including counterweights) is 34,000 tonnes.

The trials of elevator finished in July 2016, the first cargo ship was lifted in July 15, the lift time comprised 8 minutes. *Shanghai Daily* reported that the first operational use of the lift was on September 18, 2016, when limited *"operational testing"* of the lift began.

Portage Railways

Plans also exist for the construction of short portage railways bypassing the dam area altogether. Two short rail lines, one on each side of the river, are to be constructed. The 88-kilometre (55 mi) long northern portage railway will run from the Taipingxi port facility on the northern side of the Yangtze, just upstream from the dam, via Yichang East Railway Station to the Baiyang Tianjiahe port facility in Baiyang Town, below Yichang. The 95-kilometre (59 mi) long southern portage railway will run from Maoping (upstream of the dam) via Yichang South Railway Station to Zhicheng (on the Jiaozuo–Liuzhou Railway). In late 2012, preliminary work started along both future railway routes.

Relocation of Residents

Though the large size of the reservoir caused huge relocation upstream, it was considered justified by the flood protection it provides for communities downstream. As of June 2008, China relocated 1.24 million residents (ending with Gaoyang in Hubei Province) as 13 cities, 140 towns and 1350 villages either flooded or were partially flooded by the reservoir [A_2-M:CR3-1HP:S-15], about 1.5% of the province's 60.3 million and Chongqing Municipality's 31.44 million population. About 140,000 residents were relocated to other provinces.

Relocation was completed on July 22, 2008. Some 2007 reports claimed that Chongqing Municipality will encourage an additional four million people to move away from the dam to the main urban area of Chongqing by 2020. However, the municipal government explained that the relocation is due to urbanization, rather than the dam, and people involved included other areas of the municipality.

Allegedly, funds for relocating 13,000 farmers around Gaoyang disappeared after being sent to the local government, leaving residents without compensation.

Other Effects

Culture and Aesthetics

The 600 km (370 mi) long reservoir flooded some 1,300 archaeological sites and altered the appearance of the Three Gorges as the water level rose over 91 m (300 ft). Cultural and historical relics are being moved to higher ground as they are discovered, but the flooding inevitably covered undiscovered relics. Some sites could not be moved because of their location, size, or design. For example, the hanging coffins site high in the Shen Nong Gorge is part of the cliffs.

National Security

The United States Department of Defense reported that in Taiwan, "proponents of strikes against the mainland apparently hope that merely presenting credible threats to China's urban population or high-value targets, such as the Three Gorges Dam, will deter Chinese military coercion."

The notion that the military in Taiwan would seek to destroy the dam provoked an angry response from the mainland Chinese media. People's Liberation Army General Liu Yuan was quoted in the *China Youth Daily* saying that the People's Republic of China would be "seriously on guard against threats from Taiwan independence terrorists."

The Three Gorges Dam is a steel-concrete gravity dam. The water is held back by the innate mass of the individual dam sections. As a result, damage to an individual section should not affect other parts of the dam. However, damage to the entire dam through means such as missiles could cause flooding along a large area of the Yangtze River due to overflow spillage.

Structural Integrity

Days after the first filling of the reservoir, around 80 hairline cracks were observed in the dam's structure. The submerged spillway gates of the dam might pose a risk of cavitation, similar to that which severely damaged the poorly designed and cavitating spillways of the Glen Canyon Dam in the US state of Arizona, which was unable to properly withstand the Colorado river floods of 1983. However, 163,000 concrete units of the Three Gorges dam all passed quality testing and the deformation was within design limits. An experts group gave the project overall a good quality rating.

Upstream Dams

In order to maximize the utility of the Three Gorges Dam and cut down on sedimentation from the Jinsha River, the upper course of the Yangtze River, authorities plan to build a series of dams on the Jinsha, including Wudongde Dam, Baihetan Dam, along with the now completed Xiluodu and Xiangjiaba dams. The total capacity of those four dams is 38,500 MW, almost double the capacity of the Three Gorges. Baihetan is preparing for construction and Wudongde is seeking government approval. Another eight dams are in the midstream of the Jinsha and eight more upstream of it.

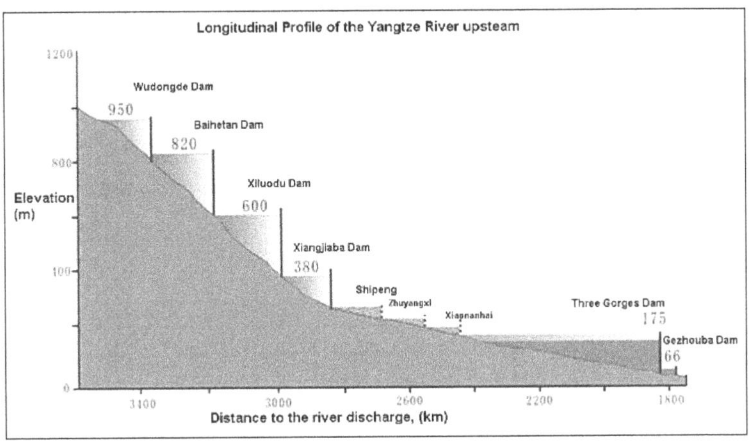

Longitudinal profile of upstream Yangtze River.

Belo Monte Dam

The Belo Monte Dam (*formerly known as* Kararaô) is a hydroelectric dam complex currently under construction on the northern part of the Xingu River in the state of Pará, Brazil. The planned installed capacity of the dam complex would be 11,233 megawatts (MW), which would make it the second largest hydroelectric dam complex in Brazil and fourth largest in the world by installed capacity, behind the Three Gorges Dam and the Xiluodu Dam in China and the Brazilian-Paraguayan Itaipu Dam. Considering the oscillations of flow river, guaranteed minimum capacity generation from the Belo Monte Dam would measure 4,571 MW, 39% of its maximum capacity.

Brazil's rapid economic growth over the last decade has provoked a huge demand for new and stable sources of energy, especially to supply its growing industries. In Brazil, hydroelectric power plants produce over 85% of the electrical energy. The Government has decided to construct new hydroelectric dams to guarantee national energy security. However, there is opposition both within Brazil and among the international community to the project's potential construction regarding its economic viability, the generation efficiency of the dams and in particular its impacts on the region's people and environment. In addition, critics worry that construction of the Belo Monte Dam could make the construction of other dams upstream- which could have greater impacts- more viable.

Plans for the dam began in 1975 but were soon shelved due to controversy; they were later revitalized in the late 1990s. In the 2000s, the dam was redesigned, but faced renewed controversy and controversial impact assessments were carried out. On 26 August 2010, a contract was signed with Norte Energia to construct the dam once the Brazilian Institute of Environment and Renewable Natural Resources (IBAMA) had issued an installation license. A partial installation license was granted on 26 January 2011 and a full license to construct the dam was issued on 1 June 2011. The licensing process and the dam's construction have been mired in federal court battles; the current ruling is that construction is allowed, because the license is based on five different environmental technical reports and in accordance with the RIMA (Environmental Impact Report, EIA-RIMA) study for Belo Monte.

The first turbines went online on 5 May 2016. As of May 2019 all turbines at Pimental and 13 turbines in main power powerhouse are online with total installed capacity of 8,176.1 megawatts (MW). The power station is planned to be completed by 2020.

Design

Overview of the dam complex.

The Belo Monte Dam (AHE Belo Monte) is a complex of three dams, numerous dykes and a series of canals in order to supply two different power stations with water. The Pimental Dam (3°27′33″S 51°57′31″W3.45917°S 51.95861°W) on the Xingu would be 36 metres (118 ft) tall and 6,248 metres (20,499 ft) long and have a structural volume of 4,768,000 cubic metres (168,400,000 cu ft). It would create the Calha Do Xingu Reservoir which would have a normal capacity of 2,069,000,000 cubic metres

(1,677,000 acre·ft) and surface area of 333 square kilometres (129 sq mi). The dam would support a power station and its spillway would serve as the complex's principal spillway with 17 floodgates and a 47,400 cubic metres per second (1,673,915 cu ft/s) maximum discharge. The dam's reservoir would also divert water into two 12 km (7 mi) long canals. These canals would supply water to the Dos Canais Reservoir, which is created within the "Big Bend" by the main dam, Belo Monte (3°06′44″S 51°48′56″W3.11222°S 51.81556°W), a series of 28 dykes around the reservoir's perimeter and the Bela Vista Dam (3°19′46″S 51°47′27″W3.32944°S 51.79083°W) which lies on the Dos Canais Reservoir's eastern perimeter. The Belo Monte Dam would support the main power station in the complex. The power station would contain twenty vertical Francis turbines listed at 550 MW (max 560 MW). Supplying each turbine with water is a 113-metre-long (371 ft), 11.2 metres (37 ft) diameter penstock, affording an average of 89.3 metres (293 ft) of hydraulic head. The Pimental Dam's power station would contain seven Kaplan bulb turbines, each rated at 25.9 MW and with 13.1 metres (43 ft) of hydraulic head.

The Belo Monte Dam would be 90 metres (300 ft) tall and 3,545 metres (11,631 ft) long and have a structural volume (embankments included) of 25,356,000 cubic metres (895,400,000 cu ft) while the Bella Vista would be 33 metres (108 ft) high and 351 metres (1,152 ft) long and have a structural volume of 239,500 cubic metres (8,460,000 cu ft). The Dos Canais Reservoir would have a normal capacity of 1,889,000,000 cubic metres (1,531,000 acre·ft), a normal surface area of 108 square kilometres (42 sq mi) and a normal elevation area of 97 metres (318 ft) above sea level. The Bela Vista Dam which serves as the complex's secondary spillway would have a maximum discharge capacity of 14,600 cubic metres per second (520,000 cu ft/s).

Power Generation and Distribution

The planned capacity of Belo Monte is listed at 11,233 MW. It is composed of the main Belo Monte Dam, and its turbine house with an installed capacity of 11,000 MW. The Pimental Dam which also includes a turbine house would have an installed capacity of 233.1 MW, containing 25.9 MW bulb turbines. The generation facility is planned to have twenty Francis turbines with a capacity of 550–611 MW each.

In February 2011, Norte Energía signed contracts with:

- IMPSA worth $450 million to design and install by the fall of 2015 four Francis turbine generation units to provide 2,500 MW of power.
- Andritz AG to provide three Francis turbines and the six bulb turbines, and the 14 excitation systems for the main power house and additional equipment for the Pimental power house.
- Alstom worth $682.3 million to provide 7 Francis turbines, and 14 gas insulated substations for the facility.

Capacity

Walter Coronado Antunes, the former Secretary of the Environment of the state of São Paulo, and ex-President of the state water and sanitation utility Sabesp, has said that the Belo Monte Dam Complex would be one of the least efficient hydro-power projects in the history of Brazil, producing only 10% of its 11,233 MW nameplate capacity between July and October (1,123 MW, and an average of only 4,419 MW throughout the year, or a 39% capacity factor). According to the president of Brazil's Energy Research Company (EPE), 39% is "just a little below" Brazil's average of 55%. Normally, the capacity factor of hydroelectric power plants is between 30% and 80%, while wind power is typically between 20% and 40%. According to a study by Eletrobras, even when at reduced capacities, Belo Monte would still have the capacity to supply the entire state of Para with electricity.

Availability

Critics claim that the project would only make financial sense if the Brazilian government builds additional dam reservoirs upstream to guarantee a year-round flow of water, thus increasing the availability of generation. Supporters of the project point out that the seasonal minimum flow of the Xingu river occurs at a time when other Brazilian hydro plants are well supplied, so that no additional dams would have to be built. Reportedly, Brazil's National Council for Power Policies approved a resolution, previously sanctioned by then president Lula, that only one hydroelectric dam would be built on the Xingu. With one dam, critics don't see an advantage regarding the dam's cost-to-benefit ratio and question why the government would just want to construct one.

Additional upstream dams would directly and indirectly affect 25,000 indigenous peoples in the entire Xingu basin. Of particular note is the Altamira (Babaquara) Dam, which would flood an additional 6,140 square kilometres (2,370 sq mi) of reservoir, according to its original design.

Developer

The project is developed by Norte Energia. The consortium is controlled by the state-owned power company Eletrobras, which directly (15%) and through its subsidiaries Eletronorte (19.98%) and CHESF (15%) controls a 49.98% stake in the consortium.

In July 2010, the federal holding company Eletrobras stated that there were 18 partners and reported their adjusted share in the project:

- Eletronorte (subsidiary of Eletrobras) – 19.98%.
- Eletrobras, state-owned company – 15%.
- CHESF (subsidiary of Eletrobras) – 15%.
- Bolzano Participacoes investments fund – 10%.

- Gaia Energia e Participações (Bertin Group) – 9%.
- Caixa Fi Cevics investments fund – 5%.
- Construction firm OAS – 2.51%.
- Queiroz Galvão, construction company – 2.51%.
- Funcef pension fund – 2.5%.
- Galvão Engenharia, construction company – 1.25%.
- Contern Construções, construction company – 1.25%.
- Cetenco Engenharia, construction company – 1.25%.
- Mendes Junior, construction company – 1.25%.
- Serveng-Civilsan, construction company – 1.25%.
- J Malucelli, construction company – 1%.
- Sinobras – 1%.
- J Malucelli Energia, construction company – 0.25%.
- SEPCO1, Siemens.

The Norte Energia consortium construction companies were reported to have originally held a 40% share.

In April 2012 it was announced that a $146 million contract was signed between Norte Energia S.A. and a consortium consisting of ARCADIS logos (a subsidiary of ARCADIS) holding a 35% share, and Themag, Concremat, and ENGECORPS, who will provide their engineering services to the project.

Economics

The dam complex is expected to cost upwards of $16 billion and the transmission lines $2.5 billion. The project is being developed by the state-owned power company Eletronorte, and would be funded largely by the Brazilian Development Bank (BNDES). The project will also include substantial amounts of funding from Brazilian pension funds Petros, Previ, and Funcef. Private investors interested in the project include mining giants Alcoa and Vale, construction conglomerates Andrade Gutierrez, Votorantim, Grupo OAS, Queiroz Galvão, Odebrecht and Camargo Corrêa, and energy companies GDF Suez and Neoenergia.

In 2006, Conservation Strategy Fund (CSF) analyzed different cost-benefit scenarios for Belo Monte as an energy project, excluding environmental costs. Initial benefits appeared marginal. When simulating energy benefits using a modeling system it became obvious that Belo Monte would require additional upstream dams to provide water storage for dry season generation. CSF concluded that Belo Monte would not be sustainable without the proposed Altamira (Babaquara) dam which would have a reservoir more than 10 times the size of Belo Monte's, flood 30 times the area submerged by Belo Monte, indigenous territories of the Araweté/Igarapé Ipixuna, Koatinemo, Arara, Kararaô, and Cachoeira Seca do Irirí natives.

Due to the project's lack of economic viability and lack of interest from private investors, the government has had to rely on pension funds and lines of credit from BNDES that draw from the Workers' Assistance Fund, oriented towards paying the public debt, to finance the project; up to one-third of the project's official cost would be financed by incentives using public money.

Alternatives

WWF-Brazil released a report in 2007 stating that Brazil could cut its expected demand for electricity by 40% by 2020 by investing in energy efficiency. The power saved would be equivalent to 14 Belo Monte hydroelectric plants and would result in national electricity savings of up to R$33 billion (US$19 billion).

Ex-director of ANEEL Afonso Henriques Moreira Santos stated that large dams such as Belo Monte were not necessary to meet the government's goal of 6% growth per year. Rather, he argued that Brazil could grow through increasing its installed capacity in wind power, currently only at 400 MW.

However, a study by the Federal University of Rio de Janeiro, published in June 2011, criticised some of these alternative suggestions and defended the Belo Monte dam project. They state that compared to the estimated costs of alternative energies, the Belo Monte dam is cheaper both in economic and in socio-environmental costs.

Environmental Effects

Chief Raoni in Paris with his petition against Belo Monte Dam.

The project is strongly criticized by indigenous people and numerous environmental organizations in Brazil plus organizations and individuals around the world.

Belo Monte's 668 square kilometres (258 sq mi) of reservoir will flood 400 square kilometres (150 sq mi) of forest, about 0.01% of the Amazon forest. Though argued to be a relatively small area for a dam's energy output, this output cannot be fully obtained without the construction of other dams planned within the dam complex. The expected area of reservoir for the Belo Monte dam and the necessary Altamira dam together will exceed 6500 km^2 of rainforest.

The environmental impact assessment written by Eletrobras, Odebrecht, Camargo Corrêa, and Andrade Gutierrez listed the following possible adverse effects:

- The loss of vegetation and natural spaces, with changes in fauna and flora.
- Changes in the quality and path of the water supply, and fish migration routes.
- Temporary disruption of the water supply in the Xingu riverbed for 7 months.

Incomplete Environmental Assessment

In February 2010, Brazilian environmental agency IBAMA granted an environmental license for the construction of the dam despite uproar from within the agency about incomplete information in the Environmental Impact Assessment (EIA) written by Eletrobras, Odebrecht, Camargo Corrêa, and Andrade Gutierrez. Previously in October 2009, a panel composed of independent experts and specialists from Brazilian universities and research institutes issued a report on the EIA, finding "various omissions and methodological inconsistencies in the EIA..." Among the problems cited within the EIA were the project's uncertain cost, deforestation, generation capacity, greenhouse gas emissions and in particular the omission of consideration for those affected by the river being mostly diverted in the 100 km (62 mi) long "Big Bend" (Volta Grande).

Two senior officials at IBAMA, Leozildo Tabajara da Silva Benjamin and Sebastião Custódio Pires, resigned their posts in 2009 citing high-level political pressure to approve the project. In January 2011, IBAMA president Abelardo Azevedo also resigned his post. The previous president Roberto Messias had also stepped down, citing in April 2010 that it was because of pressure from both the government and environmental organizations.

140 organizations and movements from Brazil and across the globe decried the decision-making process in granting the environmental license for the dams in a letter to Brazilian President Luiz Inacio Lula da Silva in 2010.

Loss of Biodiversity

The fish fauna of the Xingu river is extremely rich with an estimated 600 fish species and with a high degree of endemism. The area either dried out or drowned by the dam spans the entire known world distribution of a number of species, e.g. the zebra pleco (*Hypancistrus zebra*), the sunshine pleco (*Scobinancistrus aureatus*), the slender dwarf pike cichlid (*Teleocichla centisquama*), the killifish *Anablepsoides xinguensis* and *Spectrolebias reticulatus*, and the Xingu dart-poison frog (*Allobates crombiei*). An independent expert review of the costs of the dam concluded that the proposed flow through the Volta Grande meant the river "will not be capable of maintaining species diversity", risking "extinction of hundreds of species".

Hypancistrus zebra, one of the species found only in the Volta Grande.

Greenhouse Gas Budget

The National Amazon Research Institute (INPA) calculated that during its first 10 years, the Belo Monte-Babaquara dam complex would emit 11.2 million metric tons of Carbon dioxide equivalent, and an additional 0.783 million metric tons of CO_2 equivalent would be generated during construction and connection to the national energy grid. This independent study estimates greenhouse gas emissions of an amount that would require 41 years of optimal energy production from the Belo Monte Dam complex (including the now aborted Altamira Dam) in order to reach environmental sustainability over fossil fuel energy.

Dams in Brazil emit high amounts of methane, due to the lush jungle covered by waters each year as the basin fills. Carbon is trapped by foliage, which then decays anaerobically with help from methanogens, converting the carbon to methane, which is a more potent greenhouse gas than carbon dioxide. As a result, carbon emissions are emitted from the dam each year it is in operation. A 1990 study of the Curuá-Una Dam, also in Brazil, found that it pollutes 3.5 times more in carbon dioxide equivalent than an oil power plant generating an equal amount of electricity would; not in the form of the CO_2 atmospheric pollution associated with fossil fuel burning, but as the more dangerous methane emissions. Furthermore, the forest will be cleared before flooding of the area, so the CO_2 and methane emissions calculated for the flooding of the forested area will be significantly undercut. In addition, a study on the Brazilian Tucuruí dam showed that the actual greenhouse gas emissions were a factor ten higher than its official calculations showed, and this dam is no exception; it is feared that the Belo Monte Dam calculations are also deliberately undercutting reality and that the flooding of its reservoir will create a similar situation. If the dam builders cleared the forest beforehand, they would remove the organic matter from the reservoir floor and the dam would produce less of the greenhouse gas methane. However, in the case of Tucuruí, only the economically necessary forest was cut (10%, near the opening to the spillway) and the rest was left intact to be flooded by the reservoir. The contractors had sold the logging rights to the flooded area, but found the plot unviable in the short amount of time they had allocated before the area was set to be flooded. This forest has been decaying under the water through Methanogenesis and producing large amounts of greenhouse gases.

On the other hand, the energy generated by the dam for the next 50 years, at an average of 4419 MW, is 1.14 bboe (billion barrels of oil equivalent). This is approximately 9% of the proven oil reserves of Brazil (12.6 bbl), or 2% of the total oil reserves of Russia (60 bbl), or 5.5% of the proven oil reserves of the U.S. (21 bbl).

Social Effects

Protestors against the dam. The sign translates to “Beautiful pile of shit”.

Although strongly criticized by indigenous leaders, the president of Brazil’s EPE claims they have popular support for the dam. On 20 April 2010 *Folha de Sao Paulo* poll showed 52% in favor of the dam. The dam will directly displace over 20,000 people, mainly from the municipalities of Altamira and Vitória do Xingu. Two river diversion canals 500 metres (1,600 ft) wide by 12 kilometres (7.5 mi) long will be excavated. The canals would divert water from the main dam to the power plant. Belo Monte will flood a total area of 668 square kilometres (258 sq mi). Of the total, 400 square kilometres (150 sq mi) of flooded area will be forested land. The river diversion canals will reduce river flow by 80% in the area known as the Volta Grande (“Big Bend”), where the territories of the indigenous Juruna and Arara people, as well as those of sixteen other ethnic groups are located. While these tribes will not be directly impacted by reservoir flooding, and therefore will not be relocated, they may suffer involuntary displacement, as the river diversion negatively affects their fisheries, groundwater, ability to transport on the river and stagnant pools of water offer an environment for water-borne diseases, an issue that is criticized for not being addressed in the Environmental Impact Assessment.

Among the 20,000 to be directly displaced by reservoir flooding, resettlement programs have been identified by the government as necessary for mitigation. Norte Energia have failed to obtain free, prior, and informed consent from the Juruna and Arara indigenous tribes to be impacted by Belo Monte. The project would also attract an estimated 100,000 migrants to the area. An estimated 18,700 direct jobs would be created, with an additional 25,000 indirect jobs to accommodate the surge in population. However, only a fraction of the direct jobs will stay available after the project's completion, which critics have argued to spell economic disaster rather than economic prosperity.

The influx of immigrants and construction workers has also led to increased social tension between groups. Indigenous Groups report attacks and harassment, and in several occasions the destruction of property and the death of indigenous persons as a result from constructing and (illegal) logging activities. External researchers indicate that the majority of the Belo Monte dam's energy output will be relegated towards the aluminium industry, and will not benefit the people living in the area. However, Norte Energia released a clarification note stating their concern with the socioeconomic development of the area, including the promise to invest R$3.700 billion (1,300 million GBP) into various issues.

IBAMA Report

The IBAMA's environmental impact assessment has listed the following possible impacts:

- The generation of expectations towards the future of the local population and indigenous people.
- An increase in population and uncontrolled land occupation.
- An increase in the needs of services and goods, as well as job demand.
- A loss of housing and economic activities due to the transfer of population.
- Improvements on the accessibility of the region.
- Changes in the landscape, caused by the installation of support and main structures for the construction of the dam.
- Damage to the archaeological estates in the area.
- Permanent flooding of shelters in Gravura Assurini.

However, a clarification was released by the Brazilian authorities, in which it was deemed that the assured social and economic benefits, considered for the environmental redesign and the region's infrastructural developments, would outweigh the expected environmental damage. Since the beginning of the project many environmental and human rights organizations have been protesting against the construction of the Belo Monte Dam. On the 14th of August 2012 the Brazil Federal Court halted the construction of the Belo Monte Dam on the basis that the government's authorization of the dam was unconstitutional. The government didn't hold constitutionally required meetings with indigenous communities affected by the dam before granting permission in 2005 to start with the construction. This is against the Brazilian law and international human rights. However, Norte Energía, the company assigned with the construction of the Belo Monte Dam, has the possibility of an appeal to the Supreme Court.

Human Rights Concerns

The attitude and treatment of the Brazilian government towards the affected indigenous groups is strongly criticised internationally. The UN Human Rights Council has published statements denouncing Brazil's careless construction methods, and the International Labour Organization (ILO) likewise pointed out that the Brazilian state was in violation of ILO conventions (particularly the Indigenous and Tribal Peoples Convention, 1989 No 169) – although mechanisms for international enforcement are lacking, and it would require a Brazilian court to apply the binding principles of the convention, which Brazil has ratified. Indigenous groups have questioned the government's actions over these events, but their situation remains ignored by the authorities, as shown with the May 2011 Xingu Mission report of the CDDPH (Conselho de Defesa dos Direitos da Pessoa Humana), of which several sections regarding accusations of human right violations were excluded by the Special Secretary for Human Rights, Maria do Rosário Nunes.

Grand Coulee Dam

Grand Coulee Dam is a concrete gravity dam on the Columbia River in the U.S. state of Washington, built to produce hydroelectric power and provide irrigation water. Constructed between 1933 and 1942, Grand Coulee originally had only two powerhouses. The third powerhouse, completed in 1974 to increase energy production, makes Grand Coulee the largest power station in the United States by nameplate-capacity at 6,809 MW.

The proposal to build the dam was the focus of a bitter debate during the 1920s between two groups. One group wanted to irrigate the ancient Grand Coulee with a gravity canal while the other pursued a high dam and pumping scheme. The dam supporters won in 1933, but for fiscal reasons the initial design was for a "low dam" 290 feet (88 m) tall which would generate electricity without supporting irrigation. That year, the U.S. Bureau of Reclamation and a consortium of three companies called MWAK (Mason-Walsh-Atkinson Kier Company) began construction. After visiting the construction site in August 1934, President Franklin Delano Roosevelt endorsed the "high dam" design which, at 550 ft (168 m) high, would provide enough electricity to pump water into the Columbia basin for irrigation. Congress approved the high dam in 1935 and it was completed in 1942. The first waters overtopped Grand Coulee's spillway on June 1 of that year.

Power from the dam fueled the growing industries of the Northwest United States during World War II. Between 1967 and 1974, the third powerplant was constructed. The decision to construct the additional facility was influenced by growing energy demand, regulated river flows stipulated in the Columbia River Treaty with Canada, and competition with the Soviet Union. Through a series of upgrades and the installation of pump-generators, the dam now supplies four power stations with an installed capacity of 6,809 MW. As the centerpiece of the Columbia Basin Project, the dam's reservoir supplies water for the irrigation of 671,000 acres (2,700 km^2).

The reservoir is called Franklin Delano Roosevelt Lake, named after the United States President who presided over the dam's authorization and completion. Creation of the reservoir forced the relocation of over 3,000 people, including Native Americans whose ancestral lands were partially flooded. While the dam does not contain fish passage, neither does the next downstream dam, Chief Joseph Dam. This means no salmon reach the Grand Coulee Dam. The second large dam downstream, Rocky Reach Dam, has an intricate system of fish ladders to accommodate yearly salmon spawning and migration.

Construction

Low Dam

On July 16, 1933, a crowd of 3,000 watched the driving of the first stake at the low dam site, and excavation soon began. Core drilling commenced that September while the Bureau of Reclamation accelerated its studies and designs for the dam. It would still help control floods and provide for irrigation and hydroelectricity, though at a reduced capacity. Most importantly, it would not raise its reservoir high enough to irrigate the plateau around the Grand Coulee. However, the dam's design provided for future raising and upgrading.

The east-side cofferdam after the west base was complete.

Before and during construction, workers and engineers experienced problems. Contracts for companies to construct the various parts of the dam were difficult to award as few companies were sizable enough to fill them. This forced companies to consolidate. In addition, Native American graves had to be relocated and temporary fish ladders had to be constructed. During construction additional problems included landslides and the need to protect newly poured concrete from freezing. Construction on the downstream Grand Coulee Bridge began in May 1934 and more considerable earth-moving began in August. Excavation for the dam's foundation required the removal of 22,000,000 cu yd (17,000,000 m^3) of dirt and stone. To reduce the amount of trucking required in the excavation, a conveyor belt nearly 2 mi (3.2 km) long was built. To further secure the foundation, workers drilled 660–880 ft (200–270 m) holes into the granite and filled any fissures with grout, creating a grout curtain. At times, excavated areas collapsed from overburden. In order to secure these areas from further movement and continue excavation, 3-inch (76 mm) diameter pipes were inserted into the mass and chilled with cold liquid from a refrigeration plant. This froze the earth and secured it so construction could continue.

Final contract bidding for the dam began June 18, 1934, in Spokane, and four bids were submitted. One bid was from a lawyer with no financial backing; another was from actress Mae West which consisted of nothing more than a poem and promise to divert the river. Of the two serious bids, the lowest bid was from a consortium of three companies: Silas Mason Co. from Louisville, Kentucky, Walsh Construction Co. of Davenport, Iowa and New York and Atkinson-Kier Company of San Francisco and San Diego. The consortium was known as MWAK and their bid was $29,339,301, almost 15% lower than the $34.5 million option submitted by the next bidder, Six Companies, Inc., which was building Hoover Dam at the time.

Cofferdams

Two large cofferdams were constructed for the dam, but they were parallel to the river rather than straddling its width, so drilling into the canyon walls was not required. By the end of 1935 about 1,200 workers completed the west and east cofferdams. The west cofferdam was 2,000 ft (610 m) long, 50 ft (15 m) thick and was constructed 110 ft (34 m) above the bedrock. The cofferdams allowed workers to dry portions of the riverbed and begin constructing the dam, while water continued to flow down the center of the riverbed. In August 1936, once the west foundation was complete, portions of the west cofferdam were dismantled, allowing water to flow through part of the dam's new foundation. In February 1936, MWAK had begun constructing cofferdams above and below the channel between the east and west cofferdams. By December, the entire Columbia River was diverted over the foundations constructed within the east and west cofferdams. On December 15, 1936, the *Wenatchee Daily World* announced the river was diverted and by early the next year, people were arriving in large numbers to see the riverbed.

Design Change

Base of the dam in 1938.

On August 4, 1934, President Franklin D. Roosevelt visited the construction site and was impressed by the project and its purpose. He spoke to workers and spectators, closing with this statement: "I leave here today with the feeling that this work is well undertaken; that we are going ahead with a useful project, and we are going to see it through for the benefit of our country." Soon afterward, Reclamation was allowed to proceed with the high dam plan but faced the problems of transitioning the design and negotiating an altered contract with MWAK. In June 1935, for an additional $7 million, MWAK and Six Companies, Inc. agreed to join together as Consolidated Builders Inc. and construct the high dam. Six Companies had just finished the Hoover Dam and was nearing completion of Parker Dam. The new design, chosen and approved by the Reclamation office in Denver, included several improvements, one of which was the irrigation pumping plant.

Roosevelt envisioned the dam would fit into his New Deal under the Public Works Administration; it would create jobs, farming opportunities and would pay for itself. In addition, as part of a larger public effort, Roosevelt wanted to keep electricity prices low by limiting private ownership of utility companies, which could charge high prices for energy. Many opposed a federal takeover of the project, including its most prominent supporters, but Washington State lacked the resources to fully realize the project. In August 1935, with the help of Roosevelt and a Supreme Court decision allowing the acquisition of public land and Indian Reservations, Congress authorized funding for the upgraded high dam under the 1935 River and Harbors Act. The most significant legislative hurdle for the dam was over.

That for the purpose of controlling floods, improving navigation, regulating the flow of the streams of the United States, providing for storage and for the delivery of the stored waters thereof, for the reclamation of public lands and Indian reservations, and other beneficial uses, and for the generation of electric energy as a means of financially aiding and assisting such undertakings the projects known as "Parker Dam" on the Colorado River and "Grand Coulee Dam" on the Columbia River are hereby authorized and adopted.

First Concrete Pour and Completion

The dam after completion and water over-topping the spillway.

On December 6, 1935, Governor Clarence Martin presided over the ceremonial first concrete pour. During construction, bulk concrete was delivered on site by rail-cars where it was further processed by eight large mixers before being placed in form. Concrete was poured into 50 sq ft (4.6 m^2) columns by crane-lifted buckets, each supporting eight tons of concrete. To cool the concrete and facilitate curing, about 2,000 mi (3,200 km) of piping was placed throughout the hardening mass. Cold water from the river was pumped into the pipes, reducing the temperature within the forms from 105 °F (41 °C) to 45 °F (7 °C). This caused the dam to contract about 8 inches (20 cm) in length; the resulting gaps were filled with grout.

Until the project began, the stretch of the Columbia River where the dam was to rise was as yet unbridged, making it difficult to move men and materials. In January 1936, the Grand Coulee Bridge (a permanent highway bridge) was opened after major delays caused by high water; three additional and temporary bridges downstream had moved vehicles and workers along with sand and gravel for cement mixing. In March 1938, MWAK completed the lower dam and Consolidated Builders Inc. began constructing the high dam. The west power house was completed in December 1939 and about 5,500 workers were on site that year. Between 1940 and 1941, the dam's eleven floodgates were installed on the spillway and the dam's first generator went into operation in January 1941. The reservoir was full and the first water flowed over the dam's spillway on June 1, 1942, while work was officially complete on January 31, 1943. The last of the original 18 generators did not operate until 1950.

Reservoir Clearing

In 1933, Reclamation began efforts to purchase land behind the dam as far as 151 mi (243 km) upstream for the future reservoir zone. The reservoir, known later as Lake Roosevelt, flooded 70,500 acres (285 km^2) and Reclamation acquired an additional 11,500 acres (47 km^2) around the future shoreline. Within the zone were eleven towns, two railroads, three state highways, about one hundred and fifty miles of country roads, four sawmills, fourteen bridges, four telegraph and telephone systems, and many power lines and cemeteries. All facilities had to be purchased or relocated, and 3,000 residents were relocated. The Anti-Speculation Act was passed in 1937, limiting the amount of land farmers could own to prevent inflated prices. The government appraised the land and offered to purchase it from the affected residents. Many refused to accept the offers, and Reclamation filed condemnation suits. Members of the Colville Confederated and Spokane tribes who had settlements within the reservoir zone were also resettled. The Acquisition of Indian Lands for Grand Coulee Dam Act of June 29, 1940, allowed the Secretary of the Interior to acquire land on the Colville and Spokane Reservations, eventually accounting for 21,100 acres (85 km^2). By 1942, all land had been purchased at market value: a cost of $10.5 million that included the relocation of farms, bridges, highways and railroads. Relocation reimbursement was not offered to property owners, which was common until U.S. laws changed in 1958.

Banks and Smith felling the last tree in the reservoir zone.

In late 1938, the Works Progress Administration began clearing what would be 54,000 acres (220 km^2) of trees and other plants. The cut timber was floated downstream and sold to the highest bidder, Lincoln Lumber Company, which paid $2.25 per thousand board feet (equivalent to $40 in 2018). The pace of clearing was accelerated in April 1941 when it was declared a national defense project, and the last tree was felled on July 19, 1941. The felling was done by Reclamation Supervising Engineer Frank A. Banks and State WPA Administrator Carl W. Smith during a ceremony. 2,626 people living in five main camps along the Columbia worked on the project. When it was finished, $4.9 million had been spent in labor.

Labor and Supporting Infrastructure

Workers building the dam received an average of 80¢ an hour; the payroll for the dam was among the largest in the nation. The workers were mainly pulled from Grant, Lincoln, Douglas, and Okanogan counties and women were allowed to work only in the dorms and the cookhouse. Around 8,000 people worked on the project, and Frank A. Banks served as the chief construction engineer. Bert A. Hall was the chief inspector who would accept the dam from the contractors. Orin G. Patch served as the chief of concrete. Construction conditions were dangerous and 77 workers died.

Workers installing a penstock section.

To prepare for construction, housing for workers was needed along with four bridges downstream of the dam site, one of which, the Grand Coulee Bridge, exists today. The Bureau of Reclamation provided housing and located their administrative building at Engineer's Town, which was directly downstream of the construction site on the west side of the river. Opposite Engineer's Town, MWAK constructed Mason City in 1934. Mason city contained a hospital, post office, electricity and other amenities along with a population of 3,000. Three-bedroom houses in the city were rented for $32 a month. Of the two living areas, Engineer's City was considered to have the better housing. Several other living areas formed around the construction site in an area known as Shack Town, which did not have reliable access to electricity and the same amenities as the other towns. Incorporated in 1935, the city of Grand Coulee supported workers as well and is just west of the dam on the plateau. MWAK eventually sold Mason City to Reclamation in 1937 before its contract was completed. In 1956, Reclamation combined both Mason City and Engineer's Town to form the city of Coulee Dam. It was incorporated as a city in February 1959.

Irrigation Pumps

With the onset of World War II, power generation was given priority over irrigation. In 1943, Congress authorized the Columbia Basin Project and the Bureau of Reclamation began construction of irrigation facilities in 1948. Directly to the west and above the Grand Coulee Dam, the North Dam was constructed. This dam, along with the Dry Falls Dam to the south, enclosed and created Banks Lake, which covered the northern 27 mi (43 km) of the Grand Coulee. Additional dams, such as the Pinto and O'Sullivan Dams, were constructed alongside siphons and canals, creating a vast irrigation supply network called the Columbia Basin Project. Irrigation began between 1951 and 1953 as six of the 12 pumps were installed and Banks Lake was filled.

Expansion

Third Powerplant

Canadian Prime Minister John Diefenbaker (seated left) and US President Dwight Eisenhower at the signing of the Columbia River Treaty, 1961 After World War II, the growing demand for electricity sparked interest in constructing another power plant supported by the Grand Coulee Dam. One obstacle to an additional power plant was the great seasonality of the Columbia River's streamflow. Today the flow is closely managed—there is almost no seasonality. Historically, about 75% of the river's annual flow occurred between April and September. During low flow periods, the river's discharge was between 50,000 cu ft/s (1,400 m^3/s) and 80,000 cu ft/s (2,300 m^3/s) while maximum spring runoff flows were around 500,000 cu ft/s (14,000 m^3/s). Only nine out of the dam's eighteen generators could run year-round. The remaining nine operated for less than six months a year. In 1952, Congress authorized $125,000 for Reclamation to conduct a feasibility study on the Third Powerplant which was completed in 1953 and recommended two locations. Nine identical 108 MW generators were recommended, but as matters stood, they would be able to operate only in periods of high water.

One of the new turbines in the Third Powerplant.

Further regulation of the Columbia's flows was necessary to make the new power plant feasible. It would require water storage and regulation projects in Canada and a treaty to resolve the many economic and political issues involved. The Bureau of Reclamation and Army Corps of Engineers explored alternatives that would not depend on a treaty with Canada, such as raising the level of Flathead Lake or Pend Oreille Lake, but both proposals faced strong local opposition. The Columbia River Treaty, which had been discussed between the U.S. and Canada since 1944, was seen as the answer. Efforts to build the Third Powerplant were also influenced by competition with the Soviet Union, which had constructed power plants on the Volga River larger than Grand Coulee. On September 16, 1964, the Columbia River Treaty was ratified and included an agreement by Canada to construct the Duncan, Keenleyside, Mica Dams upstream and the U.S. would build the Libby Dam in Montana. Shortly afterward, Washington Senator Henry M. Jackson, who was influential in constructing the new power plant, announced Reclamation would present the project to Congress for appropriation and funding. To keep up with Soviet competition and increase the generating capacity it was determined the generators could be upgraded to much larger designs. With the possibility of international companies bidding on the project, the Soviets who had just installed a 500 MW hydroelectric generator on the Yenisei River indicated their interest. To avoid the potential embarrassment of an international rival building a domestic power plant, the Department of the Interior declined international bidding. The Third Powerplant was approved and President Lyndon Johnson signed its appropriation bill on June 14, 1966.

Between 1967 and 1974, the dam was expanded to add the Third Powerplant, with architectural design by Marcel Breuer. Beginning in July 1967, this involved demolishing the northeast side of the dam and building a new fore-bay section. The excavation of 22,000,000 cu yd (16,820,207 m^3) of dirt and rock had been completed before the new 1,725 ft (526 m) long section of dam was built. The addition made the original 4,300 ft (1,300 m) dam almost a mile long. Original designs for the powerhouse had twelve smaller units but were altered to incorporate six of the largest generators available. To supply them with water, six 40 ft (12 m) diameter penstocks were installed. Of the new turbines and generators, three 600 MW units were built by Westinghouse and three 700 MW units by General Electric. The first new generator was commissioned in 1975 and the final one in 1980. The three 700 MW units were later upgraded to 805 MW by Siemens.

Pump-generating Plant

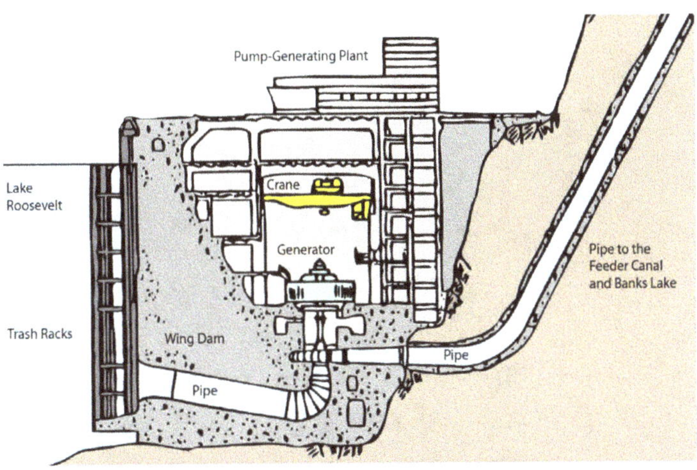

Cross-section of the Pump-Generating Plant.

After power shortages in the Northwest during the 1960s, it was determined the six remaining planned pumps be pump-generators. When energy demand is high, the pump-generators can generate electricity with water from the Banks Lake feeder canal adjacent to the dam at a higher elevation. By 1973, the Pump-Generating Plant was completed and the first two generators (P/G-7 and P/G-8) were operational. In 1983, two more generators went online, and by January 1984 the final two were operational. The six pump-generators added 314 MW to the dam's capacity. In May 2009, the Pump-Generating Plant was officially renamed the John W. Keys III Pump-Generating Power Plant after John W. Keys III, the U.S. Bureau of Reclamation's commissioner from 2001 to 2006.

Overhauls

A major overhaul of the Third Powerplant, which contains generators numbered G19 through G24, began in March 2008 and will be continuing for many years. Among the projects to be completed before the generators themselves can begin to be overhauled include replacing underground 500 kV oil-filled cables for G19, G20 and G21 generators with overhead transmission lines (started in February 2009), new 236 MW transformers for G19 and G20 (started in November 2006), and several other projects. Planning, design, procurement and site preparation for the 805 MW G22, G23 and G24 generator overhauls are scheduled to begin in 2011, with the overhauls themselves to start in 2013 with the G22 generator, then G23 starting in 2014, and finally G24 starting in 2016, with planned completions in 2014, 2016 and 2017, respectively. The generator overhauls for G19, G20 and G21 have not been scheduled as of 2010.

Operation and Benefits

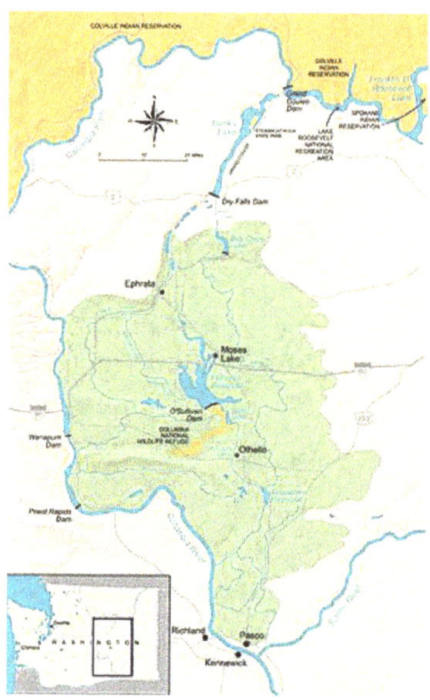

Map of the Columbia Basin Project. Green denotes land irrigated by the project. Grand Coulee Dam near top-right.

The dam's primary goal, irrigation, was postponed as the wartime need for electricity increased. The dam's powerhouse began production around the time World War II began, and its electricity was vital to the war effort. The dam powered aluminum smelters in Longview and Vancouver, Washington, Boeing factories in Seattle and Vancouver, and Portland's shipyards. In 1943, its electricity was also used for plutonium production in Richland, Washington, at the Hanford Site, which was part of the top-secret Manhattan Project. The demand for power at that project was so great that in 1943, two generators originally intended for the Shasta Dam were installed at Grand Coulee to hurry the generator installation schedule.

Irrigation

Water is pumped via the Pump-Generating Plant's 12-foot (3.7 m) diameter pipes 280 ft (85 m) from Lake Roosevelt to a 1.6 mi (2.6 km) feeder canal. From the feeder canal, the water is transferred to Banks Lake which has an active storage of 715,000 acre·ft (882 million m^3). The plant's twelve 65,000–70,000 horsepower (48,000–52,000 kW) pumps can transfer up to 1,605 cu ft/s (45 m^3/s) to the lake. Currently, the Columbia Basin Project irrigates 670,000 acres (2,700 km^2) with a potential for 1.1 million. Over 60 different crops are grown within the project and distributed throughout the United States.

Power

Grand Coulee Dam supports four different power houses containing 33 hydroelectric generators. The original Left and Right Powerhouses contain 18 main generators and the Left has an additional three service generators for total installed capacity of 2,280 MW. The first generator was commissioned in 1941 and all 18 were operating by 1950. The Third Powerplant contains a total of six main generators with a 4,215 MW installed capacity. Generators G-19, G-20 and G-21 in the Third Powerplant have a 600 MW installed capacity but can operate at a maximum capacity of 690 MW which brings the overall maximum capacity of the dam's power facilities to 7,079 MW. The Pump-Generating Plant contains six pump-generators with an installed capacity of 314 MW. When pumping water into Banks Lake they consume 600 MW of electricity. Each generator is supplied with water by an individual penstock. The largest of these feed the Third Powerplant and are 40 ft (12 m) in diameter and can supply up to 35,000 cu ft/s (990 m^3/s). The dam's power facilities originally had an installed capacity of 1,974 MW but expansions and upgrades have increased generation to 6,809 MW installed, 7,079 MW maximum. Grand Coulee Dam generates 21 TWh of electricity annually. This means the dam generates about 2,397 MW of power on average, which results in a total plant factor efficiency of 35%. In 2014, 20.24 TWh of electricity was generated.

Hydroelectric generators at Grand Coulee Dam				
Location	Type	Quantity	Capacity (MW)	Total capacity (MW)
Left Powerhouse	Francis turbine, service generator	3 (LS1-LS3)	10	30
	Francis turbine, main generator	9 (G1-G9)	125	1,125
Right Powerhouse	Francis turbine, main generator	9 (G10-G18)	125	1,125

Third Powerplant	Francis turbine, main generator	3 (G22-G24)	805	2,415
	Francis turbine, main generator	3 (G19-G21)	600 (Max: 690 MW)	1,800
Pump-Generating Plant	Pump-generator, peak generator	4 (PG9-PG12)	53.5	214
	Pump-generator, peak generator	2 (PG7-PG8)	50	100
Totals		33		6,809

Spillway

Grand Coulee Dam's spillway is 1,650 feet (500 m) long and is an overflow, drum-gate controlled type with a 1,000,000 cu ft/s (28,000 m^3/s) maximum capacity. A record flood in May and June 1948 flooded lowland below the dam and highlighted its limited flood control capability at the time, as its spillway and turbines hit a record flow of 637,800-cubic-foot-per-second (18,060 m^3/s). The flood damaged downstream riverbanks and deteriorated the face of the dam and its flip bucket at the base (toe) of the spillway. The flood spurred the Columbia River Treaty and its provisions for dams constructed upstream in Canada, which would regulate the Columbia's flow.

Pump-Generating Plant and Roosevelt Lake at bottom, feeder canal to Banks Lake at top.

Cost Benefits

The Bureau of Reclamation in 1932 estimated the cost of constructing Grand Coulee Dam (not including the Third Powerplant) to be $168 million; its actual cost was $163 million in 1943 ($1.93 billion in 2018 dollars). Expenses to finish the power stations and repair design flaws with the dam throughout the 1940s and '50s added another $107 million, bringing the total cost to $270 million ($2.06 billion in 2018 dollars), about 33% over estimates. The Third Powerplant was estimated to cost $390 million in 1967, but higher construction costs and labor disputes drove the project's final cost in 1973 to $730 million ($3.22 billion in 2018 dollars), about 55% over estimates. Despite estimates being exceeded, the dam became an economic success, particularly with the Third Powerplant exhibiting a benefit-cost ratio of 2:1. Although Reclamation has only irrigated about half of the land predicted, the gross value of crop output (in constant dollars) had doubled from 1962 to 1992, largely due to different farming practices and crop choices. The Bureau expects the money earned from supplying power and irrigation water will pay off the cost of construction by 2044.

Environmental and Social Consequences

The dam had severe negative consequences for the local Native American tribes whose traditional way of life revolved around salmon and the original shrub steppe habitat of the area. Because it lacks a fish ladder, Grand Coulee Dam permanently blocks fish migration, removing over 1,100 mi (1,770 km) of natural spawning habitat. By largely eliminating anadromous fish above the Okanogan River, the Grand Coulee Dam also set the stage for the subsequent decision not to provide for fish passage at Chief Joseph Dam (built in 1953). Chinook, Steelhead, Sockeye and Coho salmon (as well as other important species including Lamprey) are now unable to spawn in the reaches of the Upper Columbia Basin. The extinction of the spawning grounds upstream from the dam has prevented the Spokane and other tribes from holding the first salmon ceremony.

Grand Coulee Dam flooded over 21,000 acres (85 km^2) of prime bottom land where Native Americans had been living and hunting for thousands of years, forcing the relocation of settlements and graveyards. The Office of Indian Affairs negotiated with the United States Bureau of Reclamation on behalf of tribes who were concerned about the flooding of their grave sites. The Acquisition of Indian Lands for Grand Coulee Dam, 54 Stat.703 Act of June 20, 1940 allowed the Secretary of the Interior to remove human remains to new Native American grave sites. The burial relocation project started in September 1939. Human remains were put into small containers and many artifacts were discovered, but the methods of collection destroyed archaeological evidence. Various estimates for the number of relocated graves in 1939 include 915 graves reported by the Bureau of Reclamation Reclamation, or 1,388 reported by Howard T. Ball, who supervised the field work. Tribal leaders reported another 2,000 graves in 1940, but the Bureau of Reclamation would not continue grave relocation, and the sites were soon covered by water.

The town of Inchelium, Washington, home to around 250 Colville Indians, was submerged and later relocated. Kettle Falls, once a primary Native American fishing grounds, was also inundated. The average catch of over 600,000 salmon per year was eliminated. In one study, the Army Corps of Engineers estimated the annual loss was over 1 million fish. In June 1940, the Confederated Tribes of the Colville Reservation hosted a three-day event called the "Ceremony of Tears", marking the end of fishing at Kettle Falls. Within a year after the Ceremony, the falls were inundated. The town of Kettle Falls, Washington, was relocated.

Grand Coulee Dam commemorative stamp, issued 1952.

The Columbia Basin Project has affected habitat ranges for species such as mule deer, pygmy rabbits and burrowing owls, resulting in decreased populations. However, it has created new habitats such as wetlands, and riparian corridors. The environmental impact of the dam effectively ended the traditional way of life of the native inhabitants. The government eventually compensated the Colville Indians in the 1990s with a lump settlement of approximately $53 million, plus annual payments of approximately $15 million. As of 2011, the Spokane tribe still seeks compensation, currently through congressional bill H.R. 3097.

To compensate for the lack of ladder, three fisheries have been created above the dam, releasing into the upper Columbia River. One half of the fish are reserved for the displaced tribes, and one quarter of the reservoir is reserved for tribal hunting and boating.

Tourism

Built in the late 1970s, the Visitor Center contains many historical photos, geological samples, turbine and dam models, and a theater. The building was designed by Marcel Breuer, and resembles a generator rotor. Since May 1989, on summer evenings, the laser light show at Grand Coulee Dam is projected onto the dam's wall. The show includes full-size images of battleships and the Statue of Liberty, as well as some environmental comments. Tours of the Third Powerplant are available to the public and last about an hour. Visitors take a shuttle to view the generators and also travel across the main dam span (otherwise closed to the public) as the formerly used glass elevator is indefinitely out of service.

Grand Coulee Dam.

Woody Guthrie Connection

Folk singer Woody Guthrie wrote some of his most famous songs while working in the area in the 1940s. 1941 after a brief stay in Los Angeles, Guthrie and his family moved north to Oregon on the promise of a job. Gunther von Fritsch was directing a documentary about the Bonneville Power Administration's construction of the Grand Coulee Dam on the Columbia River, and needed a narrator. Alan Lomax had recommended Guthrie to narrate the film and sing songs onscreen. The original project was expected to take 12 months, but as filmmakers became worried about casting a political figure like Guthrie, they minimized his role. The Department of the Interior hired him for one month to write songs about the Columbia River and the construction of the federal dams for the documentary's soundtrack. Guthrie toured the Columbia River and the Pacific Northwest. Guthrie said he "couldn't believe it, it's a paradise", which appeared to inspire him creatively. In one month Guthrie wrote 26 songs, including three of his most famous: "Roll On, Columbia, Roll On", "Pastures of Plenty", and "Grand Coulee Dam". The surviving songs were released as *Columbia River Songs*. The film *Columbia River* was completed in 1949 and featured Guthrie's music. Guthrie had been commissioned in 1941 to provide songs for the project, but it had been postponed by WWII.

PUMPED-STORAGE HYDROELECTRIC POWER STATIONS

There's another type of hydropower plant, called the pumped-storage plant. In a conventional hydropower plant, the water from the reservoir flows through the plant, exits and is carried down stream. A pumped-storage plant has two reservoirs:

- Upper reservoir: Like a conventional hydropower plant, a dam creates a reservoir. The water in this reservoir flows through the hydropower plant to create electricity.
- Lower reservoir: Water exiting the hydropower plant flows into a lower reservoir rather than re-entering the river and flowing downstream.

Using a reversible turbine, the plant can pump water back to the upper reservoir. This is done in off-peak hours. Essentially, the second reservoir refills the upper reservoir. By pumping water back to the upper reservoir, the plant has more water to generate electricity during periods of peak consumption.

The Generator

The heart of the hydroelectric power plant is the generator. Most hydropower plants have several of these generators.

The generator, as you might have guessed, generates the electricity. The basic process of generating electricity in this manner is to rotate a series of magnets inside coils of wire. This process moves electrons, which produces electrical current.

The Hoover Dam has a total of 17 generators, each of which can generate up to 133 megawatts. The total capacity of the Hoover Dam hydropower plant is 2,074 megawatts. Each generator is made of certain basic parts:

- Shaft,
- Excitor,
- Rotor,
- Stator.

As the turbine turns, the excitor sends an electrical current to the rotor. The rotor is a series of large electromagnets that spins inside a tightly-wound coil of copper wire, called the stator. The magnetic field between the coil and the magnets creates an electric current.

In the Hoover Dam, a current of 16,500 amps moves from the generator to the transformer, where the current ramps up to 230,000 amps before being transmitted.

Bath County Pumped Storage Station

The Bath County Pumped Storage Station is a pumped storage hydroelectric power plant, which is described as the "largest battery in the world", with a maximum generation capacity of 3,003 MW, an average of 2,772 MW , and a total storage capacity of 24,000 MWh. The station is located in the northern corner of Bath County, Virginia, on the southeast side of the Eastern Continental Divide, which forms this section of the border between Virginia and West Virginia. The station consists of two reservoirs separated by about 1,260 feet (380 m) in elevation. It is the largest pumped-storage power station in the world.

Construction on the power station, with an original capacity of 2,100 megawatts (2,800,000 hp), began in March 1977 and was completed in December 1985 at a cost of $1.6 billion, Voith-Siemens upgraded the six turbines between 2004 and 2009, increasing power generation to 500.5 MW and pumping power to 480 megawatts (640,000 hp) for each turbine. Bath County Station is jointly owned by Dominion Generation (60%) and FirstEnergy (40%), and managed by Dominion. It stores energy for PJM Interconnection, a regional transmission organization in 13 states and the District of Columbia.

Design

The upper and lower reservoirs are created by earth and rock-filled embankment dams.

The upper reservoir dam is 460 feet (140 m) high, 2,200 feet (670 m) long and has a structural volume of 18,000,000 cubic yards (14,000,000 m^3). The upper reservoir on Little Back Creek has a surface area of 265 acres (107 ha) and storage capacity of 35,599 acre feet (43,911,000 m^3). The lower reservoir dam on Back Creek is 135 feet (41 m) high and 2,400 feet (730 m) in length. It has a structural volume of 4,000,000 cubic yards (3,100,000 m^3) and creates a reservoir with a surface area of 555 acres (225 ha) and storage capacity of 27,927 acre feet (34,447,000 m^3). Connecting the upper reservoir to the power station are three water conduits between 3,100–3,600 feet (940–1,100 m) long. The conduits each lead to a 990 feet (300 m) shaft which bifurcates into two penstocks (for a total of six) before reaching the turbines. Each penstock is 18 feet (5.5 m) in diameter and between 900–1,260 feet (270–380 m) in length. Maximum static pressure from the top of the water delivery system to the turbines is 1,320 feet (400 m).

Method of Operation

Water is released from the upper reservoir during periods of high demand and is used to generate electricity. What makes this different from other hydroelectric dams is that during times of low demand, power is taken from coal, nuclear, and other power plants and is used to pump water from the lower to the upper reservoir. Although this plant uses more power than it generates, it allows these other plants to operate at close to peak efficiency for an overall cost savings. Back Creek and Little Back Creek, the water sources used to create the reservoirs, have a relatively small flow rate. However, since water is pumped between the reservoirs equally, the only water taken from these creeks now that the reservoirs are full is to replace the water lost to evaporation. During operation, the water level fluctuates by over 105 feet (30 m) in the upper reservoir and 60 feet (20 m) feet in the lower reservoir.

When generating power, the water flow can be as much as 13.5 million US gallons (51,000 m^3) per minute (850 m^3/s). When storing power, the flow can be as much as 12.7 million US gallons (48,000 m^3) per minute (800 m^3/s).

Environment

A fishing habitat was created downstream of the facility. In times of drought water quality can be maintained by using nearby recreational reservoirs to supply extra water to the creeks. The creeks and recreational reservoirs have water quality sufficient for fish.

Guangdong Pumped Storage Power Station

The Guangdong Pumped Storage Power Station or Guangzhou Pumped Storage Power Station is a pumped-storage hydroelectric power station near Guangzhou, Guangdong Province, China. Power is generated by utilizing eight turbines, each with a 300 megawatts (400,000 hp) capacity, totalling the installed capacity to 2,400 megawatts

(3,200,000 hp). The generated power is sold to China Light and Power customers in Hong Kong. The power station was constructed in two stages, the first four turbines were completed in 1994 and the second four in 2000.

Operation

The station is composed mainly of a lower and upper reservoir and an underground power station. Water for the system is derived from the Liuxihe River. The lower reservoir has a capacity of 23,400,000 cubic metres (19,000 acre·ft) and is created by a 43.5 metres (143 ft) tall and 153.12 metres (502 ft) long gravity dam composed of roller-compacted concrete. Water from this reservoir is pumped into the upper reservoir which is created by a 68 metres (223 ft) tall and 318.52 metres (1,045 ft) long concrete face rock-fill embankment dam and has a capacity of 23,400,000 cubic metres (19,000 acre·ft). Water from the upper reservoir can then be re-released down two penstocks towards the power station. The same reversible pumps that moved the water up can now generate electricity. Just before reaching the power station, the two penstocks each split off into four separate branch pipes, each feeding one of the eight reversible 300 MW turbine generators with water. Once power generation is complete, the generators can reverse, pump the water back up to the upper reservoir and resume the process over again.

Ludington Pumped Storage Power Plant

The Ludington Pumped Storage Plant is a hydroelectric plant and reservoir in Ludington, Michigan. It was built between 1969 and 1973 at a cost of $315 million and is owned jointly by Consumers Energy and DTE Energy and operated by Consumers Energy. At the time of its construction, it was the largest pumped storage hydroelectric facility in the world.

Characteristics

It consists of a reservoir 110 feet (34 m) deep, 2.5 miles (4.0 km) long, and one mile (1.6 km) wide which holds 27 billion US gallons (100 Gl) or 82859 acre-feet of water. The 1.3-square-mile (3.4 km^2) reservoir is located on the banks of Lake Michigan. Because impervious bedrock is more than 800 feet (240 m) below the reservoir, the builders had to line the reservoir with a layer of asphalt and clay to prevent water seeping into the ground.

The power plant consists of six reversible turbines that can each generate 312 megawatts of electricity for a total output of 1,872 megawatts. Water is delivered from the upper reservoir to the turbines by six penstocks each 1,100 feet (340 m) long that taper from 28 to 24 feet (8.5 to 7.3 m) in diameter.

At night, during low demand for electricity, the turbines run in reverse to pump water

363 feet (111 m) uphill from Lake Michigan into the reservoir. The plant takes advantage of the natural steep sand dune landform of eastern Lake Michigan. During periods of peak demand water is released to generate power. Electrical generation can begin within two minutes with peak electric output of 1872 MW achieved in under 30 minutes. Maximum water flow is over 33 million US gallons (120,000 m^3) per minute.

This process was designed to level the load of nearby nuclear power plants on the grid. It also replaces the need to build natural gas peak power plants used only during high demand. The Ludington Pumped Storage plant is connected to six 345-kV Transmission lines, all owned and maintained by METC, a subsidiary of ITC Holdings.

The project was given the 1973 award for "Outstanding Civil Engineering Achievement" by the American Society of Civil Engineers.

Upgrade

Consumers Energy discussed plans in 2008 to extend the life of the facility and upgrade the pumps to increase efficiency by up to 9%. Consumers Energy also planned to tap the wind power resources along the eastern Lake Michigan shore with wind farms. Because wind is an intermittent power source and may inconveniently deliver large amounts of power during periods of low electric demand, pumped storage facilities are desirable to have on the same grid with large-scale wind farms. The available pumped storage capacity, along with the wind characteristics, partly determine the maximum contribution wind power can make to the overall electricity use in a region.

Consumers Energy and Detroit Edison announced an $800 million upgrade on February 7, 2011. The six-year project would begin in 2013 and extend the plant's life by at least forty years and upgrade the generating capacity from 1,872 megawatts to 2,172 megawatts. Plans for the upgrade include replacing all six of the plant's turbines, which would increase the plant's total generating capacity by 15 percent, and efficiency by 5 percent. Upon completion, the plant will produce enough power for a community of 1.65 million (a quarter million increase).

Upgradation

During its more than 40 years of operation, the massive 1,872-MW Ludington pumped-storage plant in Ludington, Mich., has pumped water from Lake Michigan to its 27 billion gallon capacity upper reservoir and generated electricity to meet peak demand. It was built from 1969 to 1973 at a cost of US$315 million. The facility, co-owned by Consumers Energy Co. and Detroit Edison, is operated by Consumers Energy and provides electricity to more than 1.3 million residential customers.

Particularly in the past few years, numerous issues have required increased maintenance, resulting in scheduled outages to perform routine maintenance every two years.

The bulk of the maintenance activity during this time was cavitation repair on the pump-turbine runners. Forced outages since the units were commissioned concern various problems with the thrust bearings. Typical observations during the planned and forced outages included:

- Wicket gate and shaft seal wear and leakage.
- Bearing and bushing wear.
- Deterioration or loss of coating on water passage.
- Stator core buckling and deterioration.
- Stator winding corona.
- Generator air cooler problems.
- Electrical and controls equipment obsolescence, including load break switches.
- Material degradation of the motor-driven exciters.

In 2006, the owners investigated the potential scope of work necessary to enable the plant's units to operate within the next 30- to 50-year period of the license to be issued in 2019 by the Federal Energy Regulatory Commission. The results of the condition assessment and life extension study performed by Devine Tarbell & Associates (DTA) (now owned by HDR) led the owners to proceed with a major overhaul and upgrade of the entire Ludington facility, beginning with the first unit.

Pump-turbine upgrade performance targets included increased turbine operating efficiency, pump operating efficiency, turbine output, pump discharge and upper reservoir elevation at which pump operation can be started. Each motor-generator was rehabilitated to enable the unit to operate reliably within its expected parameters.

Pump-turbine Design

The pump-turbines for the Ludington plant were redesigned by Toshiba using the latest technology to improve efficiency and pump cavitation performance and increase turbine output and pump discharge. The main components that needed to be replaced based on the study included the stay vanes and turbine runners. Draft tube modification was not part of this project and although Toshiba studied several wicket gate designs, the company determined new designs would not dramatically improve pump-turbine performance.

The new pump-turbine design was studied by optimization methodology combined with computational fluid dynamics (CFD) analysis. Pump-turbine hydraulic performance verification and runner blade tunings were carried out by model testing in the Swiss Federal Institute of Technology in Lausanne (EPFL-LMH).

Stay Vane Modification

The original stay vanes - 20 per unit - were designed in three different shapes for a better flow angle. The new stay vanes were modified into six different shapes, which contribute to improved efficiency in pumping and generating modes by avoiding flow separation and controlling the wake flow. The trailing edge shape of the modified stay vanes was the same, while the leading edge shapes were optimized to comply with the flow angle in generation mode at each stay vane to avoid flow separation. Modifications involved adding extension pieces to the original stay vanes' leading and trailing edges to reduce wake flow and control flow angle.

New Runners

The main goals for hydraulic performance improvement for the new runners include:

- Achieve high efficiency in both operation modes.
- Improve pump cavitation performance.
- Increase turbine output.
- Increase pump discharge.

To achieve these improvements, Toshiba applied the following technology for the new runner:

- Increased runner blades to nine from the original six.
- Deeper runner.
- Optimized blade load distributions.
- Optimized runner inlet and outlet skew angle.

By increasing the number of blades, turbine and pump efficiency at normal operating range was improved by about 4%, and it also increased pump discharge characteristics by 15%. Additionally, by using a nine-blade runner, load was reduced on each blade, leading to less cavitation.

For the new runner, the blade length is longer than a typical nine-blade runner, reducing individual blade load.

A deeper runner means the apparatus has an extended outlet (pump inlet side) on its bottom. Thus, the runner blade length is longer than a typical nine-blade runner, reducing individual blade load.

Runner inlet and outlet skew angle was optimized by CFD analysis performed by Toshiba to control secondary flow in both operating modes, especially during over- and partial-load operating conditions. Toshiba also carefully designed the runner blade outlet curve. Studying the results of cavitation model testing determined the detailed blade shape needed to provide maximum improvement of pump inlet cavitation.

Dynamic Response Study

Changing the number of blades could cause vibration from resonance due to the rotor-stator interaction between runner blades and wicket gates. Therefore, Toshiba carried out dynamic response studies in collaboration with the owner to evaluate the impact of operating using a nine-blade runner.

CFD results, runner finite element analysis (FEA) results and Toshiba's related archival data were used for the study. Results indicated:

- The expected natural frequencies for the runner in-water are far enough from exciting frequency. In addition, the expected natural frequency is higher than the exciting frequency corresponding to the rated speed, so that the exciting frequency will not coincide with the natural frequency even when increasing or decreasing speed.
- Studying vibration levels indicated a six-blade runner and nine-blade runner have the same influence on stationary parts because they both have the same exciting frequency.
- The new nine-blade runner and the stationary components will not cause any harmful vibration.

Motor-generator Design

Rehabilitating each motor-generator will enable the unit to operate reliably within its expected parameters. Rehab will also solve problems and issues that occurred with the current unit configuration, as well as uprating the units from 325 MVA/388 MVA (rated/overload) to 455 MVA based on the increased pump-turbine output. Various analyses and studies were performed to solve the problems and issues (i.e., stator core buckling, stator winding corona and thrust bearing failure). Project engineers decided to replace the entire stator, including the frame, core and windings, and thrust bearings, with improved design and manufacturing practices to alleviate these issues seen previously.

Finished, the Unit No. 1 runner assembly is a one-piece fabrication made of ASTM A487 CA6NM Class A stainless steel.

Stator coils were improved by a voltage grading system to mitigate corona activity and a copper cross section to ensure cool operation at a higher MVA output.

Stator

Because there were issues on the existing units (i.e., stator winding corona and stator core buckling, in addition to the need to increase thermal capacity of the motor-generator to 455 MVA), engineers decided to replace the stator windings and core. Toshiba performed further studies to prevent reoccurrence of stator core buckling, which is sometimes a problem with hydro generators and motor-generators designed and manufactured decades ago. In many cases, this problem is caused by insufficient clamping force on the stator core and is more likely to occur when the bottom clamping plate is an integral structure with the bottom shelf plate of the stator, as is the case at Ludington.

FEA analysis was performed on original and modified structures to verify the differences and evaluate the modified structure. As part of prudent repairs, engineers replaced the stator frame with independent bottom clamping plates.

New Design

The new stator core is constructed with high-grade, non-aging, thin silicon steel laminations. After it is machined, each lamination is coated on both sides with an insulating varnish to minimize eddy current losses.

The stator core lamination is stacked in the stator frame at the site to form a continuous core with joints between laminations overlapped. The stator core is clamped by segment-type clamping plates and fingers at both top and bottom, separated from the stator frame.

The new 20-kV stator bars have a vacuum pressure resin (VPR) rich insulation system, which consists of resin-rich mica tape around strand conductors, cured and molded by liquid medium in a vacuum pressure tank. The VPR system can reduce environmental burdens and improve working environments compared to a vacuum pressure impregnation insulation system.

The insulation performance of the new stator bars was verified by voltage endurance testing following a thermal cycle test as defined in IEEE standards 1043, 1310 and 1553.

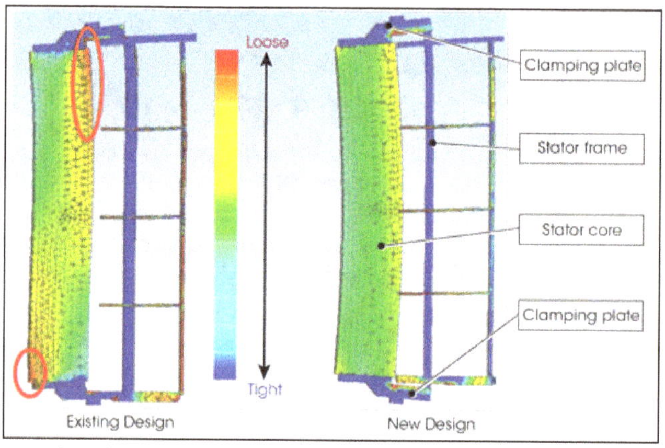

Engineers replaced the stator frame with independent bottom clamping plates.

The new stator frame is constructed of welded steel plate, split into six sections and designed for site assembly by bolting to form a complete frame. The colors in the figure show the stress distribution, not components. The new stator core is constructed with high-grade, non-aging, thin silicon steel laminations.

Thrust Bearing

At the Ludington pumped-storage plant, the existing thrust bearings had problems that required the replacement of thrust pads every couple of years. Toshiba carried out extensive analyses of the thrust bearing system, including the lower bearing bracket, to learn the cause of the problems. The data revealed:

- Excessive heat causes a large convex deformation of the thrust bearing pad in the radial direction.

- Oil film thickness around the center of the pads easily decreases against rapid thrust load change.
- There is a large difference in vertical displacement in the thrust collar between its inside and outside against thrust load, due to the collar's hollow structure.
- The lower bearing bracket is reusable with reduced stress and/or deformation by adding some reinforcements.

Project engineers determined the problems caused by the thrust collar's excessive deformation is due to its hollow structure. The excessive thrust bearing pad convex deformation is due to the relationships between the length, width and thickness of the pad. By optimizing those relationships the excessive deformation can be avoided.

In the major overhaul, the following changes and modifications were applied to the existing thrust bearing: Altering the thrust collar construction (cut the existing thrust collar from the shaft and add the new forged thrust collar to main shaft); changing to solid from hollow construction to improve the shape of the sliding surface of the thrust runner by increasing stiffness of the thrust collar; changing the thrust bearing pad construction (i.e., change its supporting system to the spring type from the pivot type); and changing L/B value (ratio of thrust pad length and width) and thickness to resolve excessive convex deformation.

Runner Manufacturing

The original runner was made of carbon steel material and had two divided halves with a crown and band cover, which covered the bolt split flanges. The new runner was made of ASTM A487 CA6NM Class A stainless steel and was designed as a one-piece fabrication. The new runner is large, about 8.4 m in diameter and weighing 270 tons, which are in the world's largest class as a pump-turbine runner.

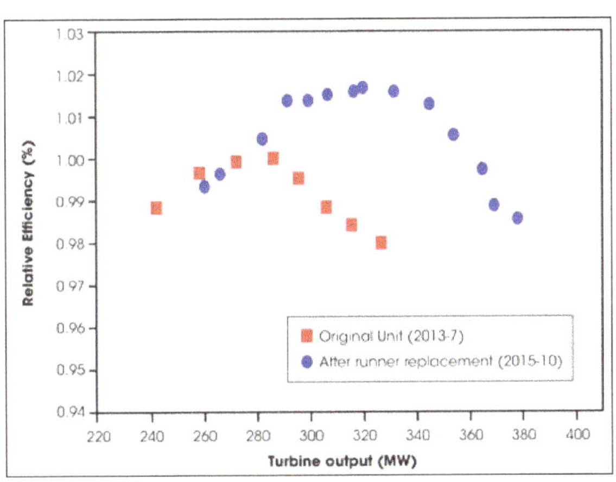

This graph indicates comparative unit relative efficiency and turbine output between the existing and new design.

One of Consumers Energy's concerns was the strength of welding seams on the fabricated runner. Consumers Energy and Toshiba decided full-penetration welding would be applied to all the new runner's welding seams to ensure sufficient strength and extended service life. To verify the strength of the prototype pump-turbine runner and the design life of 50 years, the strength and stress loaded in the runner was analyzed by Toshiba using FEA techniques and it was confirmed that the runner design meets all of the strength requirements as well as the extended service life.

Field Modifications

Stay vane extensions were installed at the site to optimize the overall hydraulic efficiency, as previously explained. The extensions were designed to fit the trailing edge and leading edge of the original stay vanes, as required. The as-found inspection revealed that the original stay vanes had cavitation damage on wide areas of both the trailing and leading edges. This affected the contact area of the stay vane extensions.

The cavitation damage on the original stay vanes had to be repaired before the extensions could be installed. The stay vane extensions consisted of three pieces: top, middle and bottom. Distortion from heat produced during welding applications was one of the risks to the stay vane and/or stay ring. The owners implemented a welding plan to minimize heat that included the procedure, material, order and stress relief.

Shaft Seal

The shaft seal modification scope included replacing the existing shaft sleeve and completely redesigning the existing shaft radial seal system. The new axial type shaft seal system includes a new rotating horizontal seal that seats against a spring and an air-loaded stationary seal. The seal also has a new cooling water system with inline filters to improve seal life, as well as provide lubrication cooling. The shaft seal modification significantly reduces water leakage through the shaft seal as compared with the existing seal.

New Performance

In 2015, operating performance of the upgraded pump-turbine was measured during performance test. Comparative performance testing utilized an eight-path ultrasonic flow meter by Caldon in the same manner used to test the original machine in 2013.

Tests confirmed turbine efficiency performance against turbine output increased to a larger output range. Stable turbine operation can be continued at 360 MW turbine capacity with 320 feet of head. The original nominal turbine capacity at the same head was 312 MW. In addition, efficiency of the new runner is the same at partial load range as the original runner.

In the case of pump operation, the pump discharge of the new runner increased about 15% compared with the original at all head ranges.

The new runner could start pumping operation at a static head of 354.1 feet, which is more than 15 feet higher than the original runner. This ability provided greater flexibility of the pumping operation and more convenient operation against grid demand.

The stator winding operation temperature met standards for Class F as defined by IEEE and 85K temperature rise, which is the industry standard for Class F.

In conclusion, unit 1's major overhaul was completed and turned to commercial operation test on March 12, 2015. The performance test carried out verified that Unit 1 has successfully achieved its target performance.

The major overhaul for Unit 2 was completed in May 2016, and Unit 3 overhaul is under way. The manufacturing for the other units is in progress and barring any unforeseen issues, all six units should be back in service by spring 2020.

RUN-OF-THE-RIVER HYDROELECTRIC POWER STATIONS

Run-of-the-river hydroelectric systems are hydroelectric systems that harvest the energy from flowing water to generate electricity in the absence of a large dam and reservoir—which is how they differ from conventional impoundment hydroelectric facilities. A small dam may be used to ensure enough water goes in the penstock, and possibly some storage (for same day use) The primary difference between this type of hydroelectric generation compared to others is that run-of-the-river primarily uses the natural flow rate of water to generate power—instead of the power of water falling a large distance. However, water may still experience some vertical drop in a run-of-the-river system from the natural landscape or small dam Another main difference between traditional hydropower is that run-of-the-river hydro is used in areas where there is little to no water storage, such as in a river.

Run-of-the-river systems like the one shown above tend to have larger hydroelectric flow rates than hydro dams that use reservoirs.

There are several classifications of run-of-the-river systems, based primarily on their capacity. The types are outlined in the table below:

Classification	Capacity
Micro	< 100 kW
Mini	100 kW - 1 MW
Small	1 - 50 MW

It is important to note that some larger scale run-of-the-river plants exist, with outputs of hundreds or thousands of MW.

Operation

For a run-of-the-river system to be possible in a given location, there needs to be two specific geographical features. The first is there must be a reasonably substantial flow rate, either from rainfall or a melting snowpack. In addition, there must be enough of a tilt to the river to speed the water up significantly. Therefore, run-of-the-river systems are best implemented in bodies of water with a fairly constant flow rate. If they are built in locations where flow rate is fairly low for a period of time and then peaks dramatically, there will be a large amount of "wasted" water during the peak flow periods as the excess water falls through the spillways. This is because these systems are built to accommodate the lowest flow rate—thus, not be able to handle significantly larger flow rates.

In run-of-the-river systems, running water from a river is guided down a channel or penstock. There can be some change in altitude at this point (from a small dam or the natural landscape) so there may still be some contribution from "falling water." The diverted water is brought to an electricity generating house. In this house, the running water drives a turbine, running a generator and generating electricity. After being used, water is fed to the river downstream.

Although run-of-the-river systems rely primarily on the flow rate of rivers to generate electricity and not a significant amount of water storage, some a small-scale dam or weir to ensure enough water enters the system itself. Pondage (a small amount of water stored behind the dam) is occasionally used, making them generally more reliable as they compensate for any discrepancies in water flow. This stored water isn't like a reservoir because it stores enough for "same day use" not future uses.

Comparison to Traditional Hydro

There are several benefits that exist as a result of using run-of-the-river hydro instead of traditional, dam based hydro. First, traditional hydro dams are expensive and time consuming to build. Comparatively speaking, run-of-the-river systems are less expensive to build and can be built over a shorter period of time. In addition, many areas where large hydro is used frequently—such as in Canada—have developed many of the existing favourable hydropower sites. Run of river systems also avoid some of the environmental problems associated with the flooding, since the pondage is much smaller than the lakes for traditional hydro.

Although there are some favourable aspects to run-of-the-river hydro, the output is significantly lower than large scale hydro. Dam based hydroelectric generation usually provides a lower cost per kWh despite the larger initial investment. Therefore, the absence of the large dam and reservoir means that the power plant will be less reliable for electricity generation. If the water levels are depleted upstream, potentially due to drought conditions, there is less water to run the hydroelectric system.

Environmental Impact

Compared to the combustion of fossil fuels, run-of-the-river hydro is responsible for fewer greenhouse gas emissions. Most of these emissions are a result of the construction of the system itself, but the operation of the plant itself contributes almost no emissions. Run-of-the-river hydroelectric plants can also be beneficial compared to impoundment dams as the small amount of water storage results in a smaller environmental footprint when compared to dams with large amounts of water storage.

Although the emission impact of these systems is less than the combustion of fuels, there are other environmental impacts that must be taken into account. First, the manipulation of river flows can cause a significant number of environmental impacts. Any diversion in the river changes how the aquatic ecosystem works, which could affect fish populations and the health of the river overall. However, the small head difference in run-of-the-river systems allows for the construction of fish ladders, which could allow fish to move through the system without being harmed or having their migration routes interrupted. As well, the changes in the river basin or water composition could increase species mortality, disrupt migration, or cause an imbalance in biodiversity. Finally, thermal pollution and increased turbidity of the exiting water are possible side effects of sending water through turbines and back into a river.

Overall, it is difficult to determine in general whether or not the damage inflicted on the environment by run-of-the-river systems is outweighed by the relatively small output as compared to large hydro dams. That means that each project must be evaluated on the specific details of the hydroelectric power plants being proposed.

Jirau Dam

The Jirau Dam is a rock-fill dam with an asphalt-concrete core, currently under construction on the Madeira River in the state of Rondônia, Brazil. The dam's hydroelectric power stations will have 50 turbines each 75 MW resulting total installed capacity of 3,750 MW. The power plant's first unit was commissioned in September 2013, the 16th in November 2014, 24th in February 2015, the 41st in December 2015, and the last in December 2016. Most of the power is designed to be exported to south-eastern Brazil via the Rio Madeira HVDC system.

The dam is part of a planned four power plant Madeira river hydroelectric complex, which will consist of two dams in Brazil (3,580 MW Santo Antonio Dam at the city of Porto Velho and Jirau), a third on the border of Brazil and Bolivia, and a fourth station inside Bolivia. Two of these, Santo Antonio and Jirau, are currently under construction, while the smaller upstream dams are still in the planning stages. In part due to the 2001–2002 power shortage in Brazil, construction of both dams was accelerated in 2009. The total estimated cost of the two facilities currently under construction is $15.6 billion ($8 billion for Jirau), including about $10 billion for the civil engineering and power plants, and $5 billion for ship locks, transmission lines, and environmental re-mediation. The Madeira river hydroelectric complex is part of the Initiative for the Integration of the Regional Infrastructure of South America, an effort by South American governments to integrate the continent's infrastructure with new investments in transportation, energy, and communication. Construction on the project was temporary halted in March 2011, February 2012 and April 2013 due to worker riots or strikes.

The Brazilian Development Bank approved an additional US$1.6 billion for the project in September 2012. The extra funding will add six more 75 MW bulb turbine-generators to the power station (a total of 50) and pay for transmission lines.

Design

The Jirau Dam is a combination embankment dam with concrete sections for the power stations and spillway. The length of the entire dam is 1,100 m (3,609 ft) while the embankment section is 800 m (2,625 ft). The embankment dam is arched, 63 m (207 ft) tall and has an asphalt-core. Its structural volume is 2,000,000 m^3 (70,629,333 cu ft) of which 17,000 m^3 (600,349 cu ft) is asphalt. The dam's spillway consists of 21 gates and has a maximum discharge of 82,000 m^3/s (2,895,803 cu ft/s). The run-of-the-river dam's power station contain 50 x 75 MW bulb turbines for a total installed capacity of 3,750 MW. The reservoir created by the dam has a surface area of 258 km^2 (100 sq mi) of which 135 km^2 (52 sq mi) is the original riverbed. Bulb turbines are a variation of the Kaplan turbine, with the main differences being that bulb turbines are installed horizontally and are generally considered to be slightly more efficient. The power plant is constructed by the French utility GDF Suez SA and Brazilian company Camargo Correa SA.

Impacts

Brazilian law requires water impoundments to undergo a very thorough approval process to ensure that each project meets environmental, social, political, and safety criteria. However, critics of the Jirau and Santo Antonio dam claim that many legal criteria were rubber-stamped before all questions from impacted groups had been addressed.

The dam's social impacts received the majority of substantive criticism. However, environmental groups noted that the fast track approval for the Madeira dams sets a dangerous precedent. Brazilian law allows for expedited licensing for eco-friendly projects described by the Worldwatch institute as "kindler, gentler dams with smaller reservoirs, designed to lessen social and environmental impacts." The Worldwatch Institute insists that no project should "fast-track the licensing of new dams in Amazonia and allow projects to circumvent Brazil's tough environmental laws".

Social

The most frequent objection is that the dam builders failed to adequately consult with indigenous peoples, as required by law. The Brazilian government indigenous protection foundation FUNAI predicts that there may be un-contacted indigenous populations in the region that will be affected by the Madeira complex. Most of the affected populations are nearest to the Jirau dam. The threat to uncontacted Indians has motivated both internal and external criticism of GDF Suez, the contractor responsible for building the Jirau dam. A coalition of non governmental organizations called for dam construction to be halted, and questions were raised during annual meeting of GDF Suez.

Moreover, federal prosecutors are suing ESBR (*Energia Sustentável do Brasil*), the company responsible for the dam, the Brazilian Institute of Environment (IBAMA) and the Brazilian Development Bank (BNDES) for the non-accomplishment of some of the conditions previewed by the environmental license and for the indemnification for losses on the traditional fishers' revenues.

Environmental

Because both the Jirau and Santo Antonio dams are run-of-the-river projects, neither dam impounds a large reservoir. Both dams also feature significant environmental re-mediation efforts. As a consequence, there has not been strong environmental opposition to the implementation of the Madeira river complex. However, critics point out that if the fish ladders fail, "several valuable migratory fish species could suffer near-extinction as a result of the Madeira dams." Jirau's environmentally friendly design earned the plant registration under the "Clean Development Mechanism" (CDM) program of the United Nations. Jirau is the largest renewable energy plant to earn the CDM, which is awarded to innovative projects that help to solve environmental problems such as climate change.

Labor

Construction on the dam was halted on 18 March 2011 as workers rioted; setting fire to buses and destroying part of the worker housing. Wages and the treatment by security officials was attributed to the rioting. Additional security personal had to be sent to the site and construction was halted. Workers went on strike at Jirau and Santo Antonio in April 2013 after a salary increase proposal was rejected.

Santo Antônio Dam

The Santo Antônio Dam is a hydroelectric dam on the Madeira River 6 km (4 mi) southwest of Porto Velho in the state of Rondônia, Brazil. The dam's run-of-the-river-hydroelectric power station have 50 turbines each rated at 71.6 MW resulting in a total installed capacity of 3,580 MW. The first unit began commercial production in March 2012, and as of June 2015 a total of 32 units were operational. Last 6 units went online in December 2016. Most of the power will be exported to south-eastern Brazil via the Rio Madeira HVDC system.

The dam is part of a planned four power plant Madeira river hydroelectric complex, which will consist of two dams in Brazil (Santo Antônio and 3,750 MW Jirau Damabout 100 km upstream), a third on the border of Brazil and Bolivia (Guayaramerin), and a fourth station inside Bolivia (Cachuela Esperanza). Santo Antonio and Jirau Dam are operating, while the smaller upstream dams are still in the planning stages. In part due to the 2001-2002 power shortage in Brazil, construction of both dams was accelerated in 2009. The total estimated cost of the two facilities currently under construction is $15.6 billion ($7 billion for Santo Antônio), including about $10 billion for the civil engineering and power plants, and $5 billion for ship locks, transmission lines, and environmental re-mediation.The Madeira river hydroelectric complex is part of the Initiative for the Integration of the Regional Infrastructure of South America, an effort by South American governments to integrate the continent's infrastructure with new investments in transportation, energy, and communication.

There are several benefits that exist as a result of using run-of-the-river hydro instead of traditional, dam based hydro. First, traditional hydro dams are expensive and time consuming to build. Comparatively speaking, run-of-the-river systems are less expensive to build and can be built over a shorter period of time. In addition, many areas where large hydro is used frequently—such as in Canada—have developed many of the existing favourable hydropower sites.Run of river systems also avoid some of the environmental problems associated with the flooding, since the pondage is much smaller than the lakes for traditional hydro.

Although there are some favourable aspects to run-of-the-river hydro, the output is significantly lower than large scale hydro. Dam based hydroelectric generation usually provides a lower cost per kWh despite the larger initial investment. Therefore, the absence of the large dam and reservoir means that the power plant will be less reliable for electricitygeneration. If the water levels are depleted upstream, potentially due to drought conditions, there is less water to run the hydroelectric system.

Opportunities for Bolivia

Bolivia has been a landlocked country since it lost its coastline to Chile in the war of the pacific in 1884. Many Bolivians feel a deep and lasting bitterness due to this loss, and the Bolivian military continues to build and maintain an open ocean navy in Lake Titicaca, awaiting an eventual recovery of access to the sea. The Madeira river complex presents an opportunity for Bolivia because all of the hydroelectric dams would feature ship locks capable of raising and lowering oceangoing vessels. If the project is completed, "more than 4,000 km of waterways upstream from the dams in Brazil, Bolivia, and Peru would become navigable." Hence, if the project is completed, both Bolivian commercial vessels and the Bolivian navy would have access to the open ocean, and lucrative sea lanes, for the first time in 120 years.

Chief Joseph Dam

The Chief Joseph Dam is a concrete gravity dam on the Columbia River, 2.4 km (1.5 mi) upriver from Bridgeport, Washington. The dam is 877 km (545 mi) upriver from the mouth of the Columbia at Astoria, Oregon. It is operated by the USACE Chief Joseph Dam Project Office, and the electricity is marketed by the Bonneville Power Administration.

Type

Chief Joseph Dam is a run-of-the-river dam which means the lake behind the dam is not able to store large amounts of water. Water flowing to Chief Joseph Dam from Grand Coulee Dam must be passed on to Wells Dam at approximately the same rate. With 27 main generators in the powerhouse, it has the hydraulic capacity of 6,030 m^3/s (213,000 cu ft/s).

In the event more water flows to Chief Joseph Dam than could be used for power generation, the spillway gates would be opened to pass the excess water. With an average annual flow rate of 3,058 m^3/s (107,992 cu ft/s), the Columbia River seldom exceeds the powerplant's capability to pass water, and spilling of water is infrequent at Chief Joseph Dam.

Reservoir

The reservoir behind the dam is named Rufus Woods Lake, and runs 82 km (51 mi) up the river channel. Bridgeport State Park, on the lake, is adjacent to the dam.

TIDAL POWER STATIONS

The gravitational pull of the moon and sun along with the rotation of the earth cause the tides. In some places, tides cause water levels near the shore to vary up to 40 feet. People in Europe harnessed this movement of water to operate grain mills more than a 1,000 years ago. Today, tidal energy systems generate electricity. Producing tidal energy economically requires a tidal range of at least 10 feet.

Tidal Barrages

One type of tidal energy system uses a structure similar to a dam called a barrage. The barrage is installed across an inlet of an ocean bay or lagoon that forms a tidal basin. Sluice gates on the barrage control water levels and flow rates to allow the tidal basin to fill on the incoming high tides and to empty through an electricity turbine system on the outgoing ebb tide. A two-way tidal power system generates electricity from both the incoming and outgoing tides.

A potential disadvantage of tidal power is the effect a tidal station can have on plants and animals in estuaries of the tidal basin. Tidal barrages can change the tidal level in the basin and increase turbidity (the amount of matter in suspension in the water). They can also affect navigation and recreation.

Several tidal power barrages operate around the world. The Sihwa Lake Tidal Power Station in South Korea has the largest electricity generation capacity at 254 Megawatts (MW). The oldest and second-largest operating tidal power plant is in La Rance, France, with 240 MW of electricity generation capacity. The next largest tidal power plant is in Annapolis Royal in Nova Scotia, Canada, with 20 MW of electricity generation capacity. China, Russia, and South Korea all have smaller tidal power plants.

Several tidal power barrages operate around the world. The Sihwa Lake Tidal Power Station in South Korea has the largest electricity generation capacity at 254 Megawatts (MW). The oldest and second-largest operating tidal power plant is in La Rance, France, with 240 MW of electricity generation capacity. The next largest tidal power plant is in Annapolis Royal in Nova Scotia, Canada, with 20 MW of electricity generation capacity. China, Russia, and South Korea all have smaller tidal power plants.

The United States does not have any tidal power plants, and it only has a few sites where tidal energy could be economical to produce. France, England, Canada, and Russia have much more potential to use tidal power.

Tidal Turbines

Tidal turbines look similar to wind turbines. They can be placed on the sea floor where there is strong tidal flow. Because water is about 800 times denser than air, tidal turbines have to be much sturdier and heavier than wind turbines. Tidal turbines are more expensive to build than wind turbines but capture more energy with the same size blades. The tidal turbine projects in Scotland and South Korea have tidal turbines with 1.5 MW electricity generation capacity. The project in Scotland is planning to have up to 400 MW of electricity generation capacity. A demonstration tidal turbine project is under development in the East River of New York.

Tidal Fences

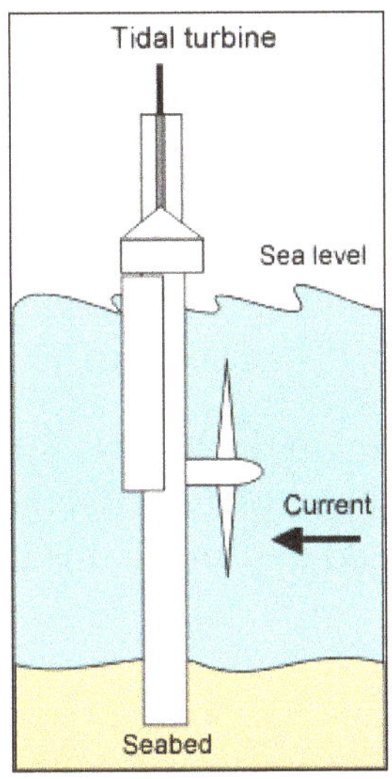

A tidal fence is a type of tidal power system that has vertical axis turbines mounted in a fence or row placed on the sea bed, similar to tidal turbines. Water passing through the turbines generates electricity.

Sihwa Lake Tidal Power Station

Sihwa Lake Tidal Power Station is the world's largest tidal power installation, with a total power output capacity of 254 MW. When completed in 2011, it surpassed the 240 MW Rance Tidal Power Station which was the world's largest for 45 years. It is operated by the Korea Water Resources Corporation.

Design

The tidal barrage makes use of a seawall constructed in 1994 for flood mitigation and agricultural purposes. Ten 25.4 MW submerged bulb turbines were driven in an unpumped flood generation scheme; power is generated on tidal inflows only, and the outflow is sluiced away, ie as one-way power generation. This slightly unconventional and relatively inefficient approach has been chosen to balance a complex mix of existing land use, water use, conservation, environmental and power generation considerations.

The station's mean operating tidal range is 5.6 m (18 ft), with a spring tidal range of 7.8 m (26 ft). The working basin area was originally intended to be 43 km^2 (17 sq mi) and has been reduced by land reclamation and freshwater dykes to 30 km^2 (12 sq mi), likely to be reduced further.

Construction

The power station was built in 2011 and started to operate in 2012. The project cost US$560 million was borne by the South Korean Government.

Environmental Context

After the seawall was built in 1994, pollution built up in the newly created Sihwa Lake reservoir, making its water useless for agriculture. Concentrations of perfluorooctane sulfonate (PFOS) measured in Lake Sihwa were among the greatest ever measured in the environment. In January 2003, PFOS had been found at 730 ng/L in Lake Shihwa water.

In 2004, seawater was reintroduced in the hope of flushing out contamination; inflows from the tidal barrage were envisaged as a complementary permanent solution. As of 2007 the power station was planned to provide this indirect environmental benefit, as well as renewable energy.

Rance Tidal Power Station

The Rance Tidal Power Station is a tidal power station located on the estuary of the Rance River in Brittany, France.

Opened in 1966 as the world's first tidal power station, it is currently operated by Électricité de France and was for 45 years the largest tidal power station in the world by installed capacity until the South Korean Sihwa Lake Tidal Power Station surpassed it in 2011.

Its 24 turbines reach peak output at 240 megawatts (MW) and average 57 MW, a capacity factor of approximately 24%. At an annual output of approximately 500 GWh (491 GWh in 2009, 523 GWh in 2010), it supplies 0.12% of the power demand of France. The power density is of the order of 2.6 W/m^2. The cost of electricity production is estimated at €0.12/kWh.

The barrage is 750 m (2,461 ft) long, from Brebis point in the west to Briantais point in the east. The power plant portion of the dam is 332.5 m (1,091 ft) long and the tidal basin measures 22.5 km^2 (9 sq mi).

Assessments

In spite of the high development cost of the project, the costs have now been recovered, and electricity production costs are lower than that of nuclear power generation (1.8 ¢/kWh versus 2.5 ¢/kWh for nuclear). However, the capacity factor of the plant is 28%, lower than 85-90% for nuclear power.

Environmental Impact

The barrage has caused progressive silting of the Rance ecosystem. Sand-eels and plaice have disappeared, though sea bass and cuttlefish have returned to the river. By definition, tides still flow in the estuary and the operator, EDF, endeavours to adjust their level to minimize the biological impact.

Tourist Attraction

The facility attracted approximately 40,000 visitors in 2011. A lock for navigation at the west end of the dam allows the passage of 1,600-tonne vessels between the English Channel and the Rance. Departmental road 168 crosses the dam and allows vehicles to travel between Dinard and Saint-Malo. There is a drawbridge where the road crosses the lock which is raised to allow larger vessels to pass. The Rance estuary is the first part of the inland waterway from the English Channel to the Bay of Biscay via the Canal d'Ille-et-Rance and the river Vilaine.

References

- Robert Ehrlich (March 13, 2013). Renewable Energy: A First Course. CRC Press. P. 219. ISBN 978-1-4665-9944-4
- Hydropower-plant1, energy, environmental: howstuffworks.com, Retrieved 2 June, 2019
- Ma, Yue (November 26, 2010). "Three Gorges Dam". Stanford University. Archived from the original on April 11, 2016. Retrieved February 13, 2016
- Major-overhaul-boosts-performance-at-ludington-pumped-storage-plant, cover-story, issue-9, volume-35: hydroworld.com, Retrieved 22 May, 2019
- "The world's largest 'L boat lift' Three Gorges Dam successfully tested" (in Chinese). Chutianjinbao News. January 14, 2016. Archived from the original on February 22, 2016. Retrieved February 15, 2016
- Hydropower-plant2, energy, environmental: howstuffworks.com, Retrieved 23 February, 2019
- Run-of-the-river-hydroelectricity, encyclopedia: energyeducation.ca, Retrieved 10 July, 2019
- Edwards-May, David (2010). Inland Waterways of France. St Ives, Cambs., UK: Imray. Pp. 90–94. ISBN 978-1-846230-14-1

6

Tidal Power

The form of hydropower that is used for the conversion of energy derived from tides into useful forms of power is referred to as tidal power. Some of the methods of generating tidal power are tidal stream generator, tidal barrage, dynamic tidal power and tidal lagoon. This chapter has been carefully written to provide an easy understanding of these methods of tidal power generation.

Tidal power, sometimes called tidal energy, is a form of hydropower that exploits the rise and fall in sea levels due to the tides, or the movement of water caused by the tidal flow. Because the tidal forces are caused by interaction between the gravity of the Earth, Moon, and Sun, tidal power is essentially inexhaustible and classified as a renewable energy source. In fact though, the ultimate energy source is the rotational energy of the Earth, which will not run out in the next four billion years, although the Earth's oceans may boil away in two billion years.

Although not yet widely used, tidal power has great potential for future electricity generation and is more predictable than wind energy and solar power. In Europe, tide mills have been used for nearly a thousand years, mainly for grinding grains.

Tidal power can be classified into two types:

- Tidal stream systems make use of the kinetic energy from the moving water currents to power turbines, in a similar way to underwater wind turbines. This method is gaining in popularity because of the lower ecological impact compared to the second type of system, the barrage.
- Barrages make use of the potential energy from the difference in height (or *head*) between high and low tides, and their use is better established. These suffer from the dual problems of very high civil infrastructure costs and environmental issues.

Modern advances in turbine technology may eventually see large amounts of power generated from the oceans using the tidal stream designs. Arrayed in high velocity areas where natural flows are concentrated, such as the west coast of Canada, the Strait of Gibraltar, the Bosporus, and numerous sites in south east Asia and Australia. Such flows occur almost anywhere where there are entrances to bays and rivers, or between land masses where water currents are concentrated.

A factor in human settlement geography is water. Human settlements have often started around bays, rivers, and lakes. Future settlement may be concentrated around moving water, allowing communities to power themselves with non-polluting energy from moving water.

Barrage Tidal Power

Rance tidal power plant.

An artistic impression of a tidal barrage, including embankments, a ship lock and caissons housing a sluice and two turbines.

The barrage method of extracting tidal energy involves building a barrage and creating a tidal lagoon. The barrage traps a water level inside a basin. Head (a height of water pressure) is created when the water level outside of the basin or lagoon changes relative to the water level inside. The head is used to drive turbines. The largest such installation has been working on the Rance river, France, since 1967, with an installed (peak) power of 240 MW, and an annual production of 600 GWh (about 68 MW average power).

The basic elements of a barrage are caissons, embankments, sluices, turbines, and ship locks. Sluices, turbines, and ship locks are housed in caisson (very large concrete blocks). Embankments seal a basin where it is not sealed by caissons.

The sluice gates applicable to tidal power are the flap gate, vertical rising gate, radial gate and rising sector.

Barrage systems are sometimes affected by problems of high civil infrastructure costs associated with what is in effect a dam being placed across two estuarine systems, and the environmental problems associated with changing a large ecosystem.

Ebb Generation

The basin is filled through the sluices until high tide. Then the sluice gates are closed. (At this stage, there may be "Pumping" to raise the level further.) The turbine gates are kept closed until the sea level falls to create sufficient head across the barrage, and then are opened so that the turbines generate until the head is again low. Then the sluices are opened, turbines disconnected and the basin is filled again. The cycle repeats itself. Ebb generation (also known as outflow generation) takes its name because generation occurs as the tide ebbs.

Flood Generation

The basin is filled through the turbines, which generate at tide flood. This is generally much less efficient than ebb generation, because the volume contained in the upper half of the basin (which is where ebb generation operates) is greater than the volume of the lower half (and making the difference in levels between the basin side and the sea side of the barrage), less than it would otherwise be. This is not a problem with the "lagoon" model; the reason being that there is no current from a river to slow the flooding current from the sea.

Pumping

Turbines are able to be powered in reverse by excess energy in the grid to increase the water level in the basin at high tide (for ebb generation). This energy is more than returned during generation, because power output is strongly related to the head.

Two-basin Schemes

With two basins, one is filled at high tide and the other is emptied at low tide. Turbines are placed between the basins. Two-basin schemes offer advantages over normal schemes in that generation time can be adjusted with high flexibility and it is also possible to generate almost continuously. In normal estuarine situations, however, two-basin schemes are very expensive to construct due to the cost of the extra length of barrage. There are some favorable geographies, however, which are well suited to this type of scheme.

Environmental Impact

The placement of a barrage into an estuary has a considerable effect on the water inside the basin and on the fish. A tidal current turbine will have a much lower impact.

Turbidity

Turbidity (the amount of matter in suspension in the water) decreases as a result of smaller volume of water being exchanged between the basin and the sea. This lets light from the Sun to penetrate the water further, improving conditions for the phytoplankton. The changes propagate up the food chain, causing a general change in the ecosystem.

Salinity

As a result of less water exchange with the sea, the average salinity inside the basin decreases, also affecting the ecosystem. "Tidal Lagoons" do not suffer from this problem.

Sediment Movements

Estuaries often have high volume of sediments moving through them, from the rivers to the sea. The introduction of a barrage into an estuary may result in sediment accumulation within the barrage, affecting the ecosystem and also the operation of the barrage.

Pollutants

Again, as a result of reduced volume, the pollutants accumulating in the basin may be less efficiently dispersed, so their concentrations may increase. For biodegradable pollutants, such as sewage, an increase in concentration is likely to lead to increased bacteria growth in the basin, having impacts on the health of the human community and the ecosystem.

Fish

Fish may move through sluices safely, but when these are closed, fish will seek out turbines and attempt to swim through them. Also, some fish will be unable to escape the water speed near a turbine and will be sucked through. Even with the most fish-friendly turbine design, fish mortality per pass is approximately 15 percent (from pressure drop, contact with blades, cavitation, and so forth.). This can be acceptable for a spawning run, but is devastating for local fish who pass in and out of the basin on a daily basis. Alternative passage technologies (fish ladders, fish lifts, and so forth) have so far failed to solve this problem for tidal barrages, either offering extremely expensive solutions, or ones which are used by a small fraction of fish only. Research in sonic guidance of fish is ongoing.

Tidal Stream Power

A relatively new technology tidal stream generators draw energy from currents in much the same way as wind turbines. The higher density of water, some 832 times the density of air, means that a single generator can provide significant power.

Even more so than with wind power, selection of location is critical for a tidal stream power generator. Tidal stream systems need to be located in areas with fast currents where natural flows are concentrated between obstructions, for example at the entrances to bays and rivers, around rocky points, headlands, or between islands or other land masses.

Prototypes

Several commercial prototypes have shown promise. Trials in the Strait of Messina, Italy, shave been carried out successfully, and an Australian company, Tidal energy, undertook successful commercial trials of highly efficient shrouded turbines on the Gold Coast, Queensland, in 2002, that was followed by successful joint venture commercial trials by Canada by Quantum Hydro Power in 2005-2006, using the Gorlov Helical Turbine on the Canadian West Coast where water speeds have been measured up to 16 knots.

During 2003, a 300 kW Periodflow marine current propeller type turbine was tested off the coast of Devon, England, and a 150 kW oscillating hydroplane device, the Stingray, was tested off the Scottishcoast.

Energy Calculations

The energy available from these kinetic systems can be expressed as:

$$P = Cp \times 0.5 \times \rho \times A \times V^3$$

Where,

Cp is the turbine coefficient of performance.

P = the power generated (in kW).

ρ = the density of the water (seawater is 1025 kg per cubic meter (CM)).

A = the sweep area of the turbine (in m^2).

V^3 = the velocity of the flow cubed (that is, V x V x V).

Relative to an open turbine in free stream where typical turbine efficiency has been measured at around 22 percent. Shrouded turbines are capable of higher efficiencies as much as 4 times the power of the same turbine in open flow.

Shrouded Turbines

A shroud or duct that houses the turbine and can produce up to four times the energy output of the same turbine for little extra project cost has been proven by prototype tests in Australia in 2002 and 2005.

Considered to be the next generation of water turbines, they are an advancement over technology that has been relatively stagnant since the middle ages when the windmill was invented.

Shrouded turbines can produce the same energy from a much smaller turbine while larger turbines if housed in a shroud have significant attraction to commerce where “Rate Of Return” on investment may be up to fifteen years on some sites where open turbines are used.

Commercial trials undertaken on shrouded turbines in 2006, along the Canadian west coast will eventually see mass production and roll out over multiple sites where there are some of the fastest flows ever recorded of up to 16 knots, or about 8.5 m/s. A one meter (m) square open turbine in this flow could produce around 0.5 megawatt (MW), while a shrouded turbine nearly 2 MW. Average flows are somewhat less around 6 to 10 knots or about 3-5 m/s.

Variable Nature of Power Output

Tidal power schemes do not produce energy all day. A conventional design, in any mode of operation, would produce power for 6 to 12 hours in every 24 and will not produce power at other times. As the tidal cycle is based on the rotation of the Earth with respect to the moon (24.8 hours), and the demand for electricity is based on the period of rotation of the earth (24 hours), the energy production cycle will not always be in phase with the demand cycle. However, the tides are relatively reliable and more predictable than other alternative energy sources, such as wind.

Mathematical Modeling of Tidal Schemes

In mathematical modeling of a scheme design, the basin is broken into segments, each maintaining its own set of variables. Time is advanced in steps. Every step, neighboring segments influence each other and variables are updated.

The simplest type of model is the flat estuary model, in which the whole basin is represented by one segment. The surface of the basin is assumed to be flat, hence the name. This model gives rough results and is used to compare many designs at the start of the design process.

In these models, the basin is broken into large segments (1D), squares (2D), or cubes (3D). The complexity and accuracy increases with dimension.

Mathematical modeling produces quantitative information for a range of parameters, including:

- Water levels (during operation, construction, extreme conditions, and so forth).
- Currents.

- Waves.
- Power output.
- Turbidity.
- Salinity.
- Sediment movements.

Energy Efficiency

Tidal energy has an efficiency of 80 percent in converting the potential energy of the water into electricity. It is thus very efficient, compared to other energy resources such as solar power or fossil fuel power plants.

Global Environmental Impact

A tidal power scheme is a long-term source of electricity. A proposal for the Severn Barrage, if built, has been projected to save 18 million tons of coal per year of operation. This decreases the output of greenhouse gases into the atmosphere.

If fossil fuel resource is likely to decline during the twenty-first century, as predicted by Hubbert peak theory, tidal power is one of the alternative source of energy that will need to be developed to satisfy the human demand for energy.

TIDAL STREAM GENERATOR

Tidal stream generators make use of the kinetic energy of moving water to turn a turbine—similar to the way a wind turbine uses wind to create electricity. However, the power available for tidal power generation in a given area can be greater than a wind turbine due to the higher density of water. These types of tidal generators tend to be the cheapest (though still quite expensive) and most environmentally friendly of any type of tidal power generation. There are several different specific types of tidal stream generators.

These types of generators have a very low visual impact and are mostly or totally submerged. Furthermore, they are less intrusive to marine life because they produce less noise pollution, tides are more predictable than wind (which varies a great deal). The density of water is also much higher than air, which means that tidal turbines can be much smaller than wind turbines for the same output. This is further enhanced by the use of the Venturi effect, which is a way of getting water to move faster through these turbines.

The initial investment costs for all types of tidal power technology are extremely high and it takes quite a while to make back that investment. In addition, sea water is quite corrosive which leads to high maintenance costs.

Since the tides flow-in and recede once per day, tidal stream generators provide intermittent power generation.

Varying Types of Generators

Different designs of tidal power generation may prove to offer different advantages, extensive studies are still needed on this relatively new technology. Studies are needed for maintenance costs vs. income and environmental impactsover an extended period of time. Power output can vary considerably between these classes of turbines. However, since the tidal power industry is so new predictions for power output vary considerably. Currently, research data on stream generators is scarce due to the emerging nature of the industry.

Axial (Horizontal) Turbines

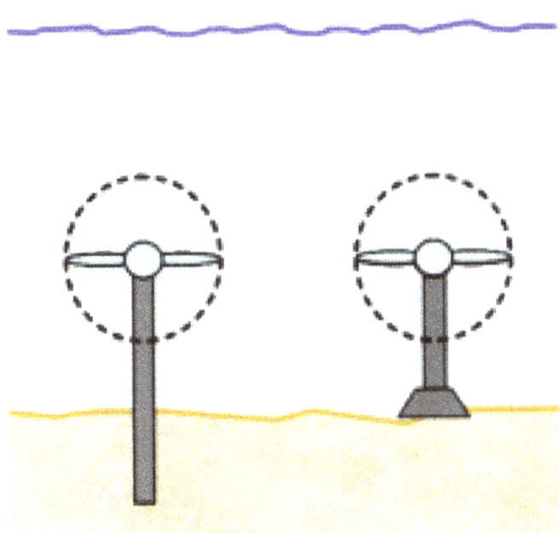

Axial turbines underwater.

Axial turbines have a rotor that is parallel to the incoming water stream. They use rotors similar to those seen on wind turbines but have modifications due to the differing fluid properties of water from air.

These devices use the lift of water to generate power. This requires specially shaped airfoil surfaces designed to create a pressure difference. This leads to a net force in the direction perpendicular to the water flow and thus turns the device. Rotors of this type must be carefully oriented (the orientation is referred to as the rotor pitch), to maintain their ability to harness the power of the tide as it changes.

Cross-flow Turbines

Cross flow turbines have a rotor that is perpendicular to the water flow but parallel to the water surface, Unlike axial turbines, these devices typically use the *drag* of water to generate power. In drag-based turbines, the force of the water pushes against a surface, like wind on an open sail. This works because the drag of the open face of the turbine blade is greater than the drag on its closed face. Drag based devices are inherently less efficient than their lift-based counterparts because they operate with respect to the relative flow speed of the water. This means that as a turbine rotates faster the relative flow speed of the water will decrease and thus transfer less of its energy.

A cross flow water turbine.

Vertical Axis Turbines

With vertical axis turbines the rotational axis of the rotor is vertical to the water surface and also perpendicular to the incoming water flow. Both lift and drag type blades can be used in these turbines.

Oscillating Generators

Oscillating devices do not have a rotational component, instead these generators make use of sections which are pushed sideways by the flowing current to create a hydraulic pump. This pump transfers its energy to a motor, which then turns a generator and creates electricity.

Tidal Stream Developers

There are a number of individuals and companies developing tidal energy converters across the world.

Tidal Stream Testing

The world's first marine energy test facility was established in 2003 to kick start the development of the wave and tidal energy industry in the UK. Based in Orkney, Scotland, the European Marine Energy Centre (EMEC) has supported the deployment of more wave and tidal energy devices than at any other single site in the world. EMEC provides a variety of test sites in real sea conditions. Its grid connected tidal test site is located at the Fall of Warness, off the island of Eday, in a narrow channel which concentrates the tide as it flows between the Atlantic Ocean and North Sea. This area has a very strong tidal current, which can travel up to 4 m/s (8 knots) in spring tides. Tidal energy developers currently testing at the site include Alstom (formerly Tidal Generation Ltd), andritz hydro Hammerfest, OpenHydro, Scotrenewables Tidal Power, and Voith.

Commercial Plans

RWE's npower announced that it is in partnership with Marine Current Turbines to build a tidal farm of SeaGen turbines off the coast of Anglesey in Wales, near the Skerries.

"The Skerries project located in Anglesey, Wales, will be one of the first arrays deployed using the Siemens owned Marine Current Turbines SeaGen S tidal turbines. The marine consent for the project was recently awarded, the first tidal array to be consented in Wales. The 10MW array will be fully operational in 2015." - CEO of Siemens Energy Hydro & Ocean Unit Achim Wörner.

In November 2007, British company Lunar Energy announced that, in conjunction with E.ON, they would be building the world's first deep-sea tidal energy farm off the coast of Pembrokeshire in Wales. It will provide electricity for 5,000 homes. Eight underwater turbines, each 25 metres long and 15 metres high, are to be installed on the sea bottom off St David's peninsula. Construction is due to start in the summer of 2008 and the proposed tidal energy turbines, described as "a wind farm under the sea", should be operational by 2010.

British Columbia Tidal Energy Corp. plans to deploy at least three 1.2 MW turbines in the Campbell River or in the surrounding coastline of British Columbia by 2009.

Alderney Renewable Energy Ltd was granted a licence in 2008 and is planning to use tidal turbines to extract power from the notoriously strong tidal races around Alderney in the Channel Islands. It is estimated that up to 3 GW could be extracted. This would not only supply the island's needs but also leave a considerable surplus for export, using a France-Alderney-Britain cable (FAB Link) which is expected to go online by 2020.

Nova Scotia Power has selected OpenHydro's turbine for a tidal energy demonstration project in the Bay of Fundy, Nova Scotia, Canada and Alderney Renewable Energy Ltd for the supply of tidal turbines in the Channel Islands.

Pulse Tidal are designing a commercial device with seven other companies who are expert in their fields. The consortium was awarded an €8 million EU grant to develop the first device, which will be deployed in 2012 and generate enough power for 1,000 homes.

ScottishPower Renewables are planning to deploy ten 1MW HS1000 devices designed by Hammerfest Strom in the Sound of Islay.

In March 2014, the Federal Energy Regulatory Committee (FERC) approved a pilot license for Snohomish County PUD to install two OpenHydro tidal turbines in Admiralty Inlet, WA. This project is the first grid-connected two-turbine project in the US; installation is planned for the summer of 2015. The OpenHydro tidal turbines that Snohomish County PUD will use are designed to be placed directly into the seafloor at a depth of roughly 200 feet, so that there will be no effect on commercial navigation overhead. The license granted by the FERC also includes plans to protect fish, wildlife, as well as cultural and aesthetic resources, in addition to navigation. Each turbine measures 6 meters in diameter, and will generate up to 300 kW of electricity. In September 2014, the project was canceled due to cost concerns.

Energy Calculations

Turbine Power

Tidal energy converters can have varying modes of operating and therefore varying power output. If the power coefficient of the device CP" is known, the equation below can be used to determine the power output of the hydrodynamic subsystem of the machine. This available power cannot exceed that imposed by the Betz limit on the power coefficient, although this can be circumvented to some degree by placing a turbine in a shroud or duct. This works, in essence, by forcing water which would not have flowed through the turbine through the rotor disk. In these situations it is the frontal area of the duct, rather than the turbine, which is used in calculating the power coefficient and therefore the Betz limit still applies to the device as a whole.

The energy available from these kinetic systems can be expressed as:

$$P = \frac{\rho A V^3}{2} C_P$$

where,

C_P" = the turbine power coefficient.

P = the power generated (in watts).

ρ = the density of the water (seawater is 1027 kg/m^3).

A = the sweep area of the turbine (in m^2).

V = the velocity of the flow.

Relative to an open turbine in free stream, ducted turbines are capable of as much as 3 to 4 times the power of the same turbine rotor in open flow.

Resource Assessment

While initial assessments of the available energy in a channel have focus on calculations using the kinetic energy flux model, the limitations of tidal power generation are significantly more complicated. For example, the maximum physical possible energy extraction from a strait connecting two large basins is given to within 10% by:

$$P = 0.22 \rho g \Delta H_{max} Q_{max}$$

where,

ρ = the density of the water (seawater is 1027 kg/m^3).

g= gravitational acceleration (9.80665 m/s^2).

ΔH_{max} = maximum differential water surface elevation across the channel.

Q_{max} = maximum volumetric flow rate though the channel.

Environmental Impacts

The main environmental concern with tidal energy is associated with blade strike and entanglement of marine organisms as high speed water increases the risk of organisms being pushed near or through these devices. As with all offshore renewable energies, there is also a concern about how the creation of EMF and acoustic outputs may affect marine organisms. Because these devices are in the water, the acoustic output can be greater than those created with offshore wind energy. Depending on the frequency and amplitude of sound generated by the tidal energy devices, this acoustic output can have varying effects on marine mammals (particularly those who echolocate to communicate and navigate in the marine environment such as dolphins and whales). Tidal energy removal can also cause environmental concerns such as degrading farfield water quality and disrupting sediment processes. Depending on the size of the project, these effects can range from small traces of sediment build up near the tidal device to severely affecting nearshore ecosystems and processes.

One study of the Roosevelt Island Tidal Energy (RITE, Verdant Power) project in the East River (New York City), used 24 split beam hydroacoustic sensors (scientific echosounder) to detect and track the movement of fish both upstream and downstream of each of six turbines. The results suggested (1) very few fish using this portion of the river, (2) those fish which did use this area were not using the portion of the river which would subject them to blade strikes, and (3) no evidence of fish traveling through blade areas.

Work is currently being conducted by the Northwest National Marine Renewable Energy Center (NNMREC) to explore and establish tools and protocols for assessment of physical and biological conditions and monitor environmental changes associated with tidal energy development.

Tidal Farm

A tidal farm is a group of multiple tidal stream generators assembled in the same location used for production of electric power, similar to that of a wind farm. The low-voltage powerlines from the individual units are then connected to a substation, where the voltage is stepped up with the use of a transformer for distribution through a high voltage transmission system.

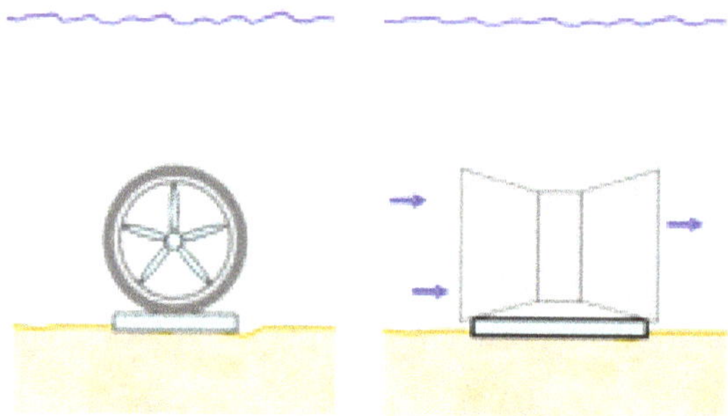

Sea Bed Turbine.

A mathematical optimization approach is used design turbine farm layouts. Using the environmental parameters such as water depth and incorporating them using mathematical formulas, a farm layout can be developed and tested. Through this research and development, factors such as the number of turbines, location of turbines and overall farm profit could be accurately tested and predicted.

In Brittany, France, a French tidal farm has deployed its first of two 500 KW turbines. The Project is located 16 miles offshore and has a depth of 35 meters. Once the power is generated it will be converted and transported to an onshore site located in the Arcouest Peninsula in Ploubazlanec. This project contributes towards the progress in a shifts towards renewable energy and tidal energy in particular.

In Iran there have been interest in development of tidal stream energy due to its predictability, and consistency. Faculty of Civil Engineering at the university of Tarbiat Modares University have identified potential sites of interest of these farms. Locations include the Persian Gulf, the Oman Sea, Khowran Straight, Hengam Island and Greater Tunb Island. Cost of energy, power output, Tariff rates and expected return on investments have also been factors analyzed and studied by the university research group.

Methods have been developed in order to determine the most suitable sites for tidal farms. Factors considered in the selection of locations include potential to provide the greatest power performance, the least amount of cost, and the least amount of inconvenience towards marine life. A case study done in the Bristol Channel used a hydrodynamic model on a Matlab-based program in order to obtain its results.

Technology

Tidal Farms utilize tidal stream generators that are grouped together to produce electricity. These generators use the moving tides to turn turbines that are very similar the wind turbines used on land. The power of the ocean and the turbines advance technology guarantee a much more predictable energy output then regular wind turbines. The turbines are usually located in areas with high tidal activity in order for the generators to be as efficient as possible. What makes tidal farms unique is that they are set up in groups to allow much more energy production. The generators are connected to substations on shore to transform voltage from high to low, or low to high. These generators can be semi-submersible or fixed into the sea floor, which means they would be out of sight and not an eyesore for the public. The turbines that would be used would be very slow moving due to the density of the water, this is very beneficial to the aquatic life because fish would be able to freely pass through without being in danger of dying. Some turbines can also be used in irrigation canals, rivers, and dam whether the flow of water is fast or slow.

Operators

Scotland is one of the main leaders in the effort to utilize tidal energy as an alternative energy resource. In 2012, Scottish Power installed a 30 ft. turbine off of the Orkney Islands. The currents off of these islands are very fast moving and the tests conducted had shown that the generator produced one megawatt of electricity, enough to power 500 homes. Scotland is also looking to install a more powerful generator off of the Sound of Islay that would be capable of powering in upwards of 5,000 homes once fully operational. In January 2015, production of a 400-megawatt tidal generator was being constructed in Northern Scotland. This generator would be capable of powering 175,000 homes. Ocean power is a clean and efficiency with an energy source that never turns off. Using Tidal farms is a much cleaner and efficient way to produce electricity. One of the drawbacks to tidal farms is marine life and how it will affect it. They would also need to set the tidal farms deep in the ocean where it won't affect fishing boats or large ship passing by. The United States of America has nearly 12,380 miles of coastline, and is undergoing an 18-month study to see how well the tidal farms work *source*. The project could cost up to $10 million, including $2 million on fish monitoring equipment and if everything goes to plan the U.S. could be seeing tidal farms along the US coast. The zero-emission tidal farms could be the way to a cleaner planet, a better future and maybe reducing the cost of electricity in the long run.

Leaders in Practice

The company Ocean Flow used Siemens technology to design a semi-submersible turbine. The models created have proven to be able to withstand sea conditions excellently. The company claims its model creates less disturbances to its surrounding ecosystem, and cost less to deploy. A key feature to this model set up is the platform that the turbine rest on. It was developed at Newcastle University School of Marine Science and Technology in 2006. The Platform is suitable to withstand harsh conditions in deep depth of the ocean. Senior development engineer at OceanFlow Mark Knos commented on the project and stated that they had created a 1/40th scale model to test and a 1/10 scale model as well. Both models were tested and have provided promising results.

Types

- Double and Three Bladed Turbine are turbines attached to a stationary pole and rotate axial to chase the ocean currents. Some of the double and triple bladed turbines can have two sets attached to the pole for better efficiency. This type of turbine has to be detached from the stationary pole and lifter with cranes attached to ships when maintenance is to be performed.
- Semi Submersible Turbines is a more expensive turbine but in the long run it is cheaper and is more cost efficient. The turbines are connecting to a stationary post and the turbine generator can be raised and lowered anytime for maintenance.
- Duct Style Turbine uses duct all the way around the entrance of the turbine to guide and accelerate the tidal stream toward the rotor. By using a duct, more energy can be extracted from the same amount of water with smaller diameter rotor blades thereby keeping costs of manufacture and maintenance down.
- Cable Tethered Turbine floating turbines are attached by a chain to a stationary point in the bottom of the ocean and follow the ocean current in a horizontal 360 degrees. These turbines are easy to bring up to the surface for maintenance because they are pressurized with air. They are also equipped with sensors to detect any water trying to make its way into the pressurized generator.

Problems

One of the few environmental unknowns about tidal farms is the threat they may pose to the plant life in the areas that the turbines would be set up. But by having the blades turn at a slower than normal speed wind turbines can eliminate some of the potential environmental problems. Another problem that can occur is making the turbines water tight to prevent seawater from corrosion the metal parts inside the turbine. Underwater turbines would have to be position away from shipping lanes, too close to shore, and in deep enough waters for them not to interfere with everyday shipping traffic. Having countries like Scotland position underwater turbines can help other countries learn and explore better ideas for creating energy by learning from the successes and the failures achieved by the leading countries. Scotland expects the rest of the world to follow their example and install tidal farms all over the world in the efforts to help stop pollution and make producing energy cleaner and safer.

Effects on Ocean Geography and Tidal Patterns

Tidal farms present many possible environmental and ecological changes to the environment they are placed in. The structures of these farms produce changes in tidal patterns, sediment flow, and water column turbulence. Tidal patterns can be affected in different ways depending on the structure of the farm itself. The structure is referring to the size and cross-sectional area because the number of blades or the load that is put on the farm doesn't make much of a difference on the overall height of the water. These effects can be seen to lower both the high tide and low tides which means that the water level will be lower in total. The number of blades and load on the system however does affect the tidal range vastly. Tidal range can be reduced up to 42% with observations of the densest farms which can cause destruction to 32% of the areas around them, but it can be controlled. Damage can be reduced to 19% by using a two-rotor spaced turbine and only 5.4% with five-rotor spaced turbines. This is done by spreading the water's work over a large surface area on the turbine itself to mitigate the alteration of the flow in the water column.

Sediment flow is affected by the introduction of tidal farms to an area. Not only does it create turbulence that moves the sediment, but it changes the ecosystem around it. By moving more sediment into areas that didn't gain much before, places like grass beds could be wiped out by being covered in sediment. The geography on the ocean floor would be changed by the new sediment flow patterns because of the new turbulence. New things like sandbars could form around the farms causing even more of an impact to the surrounding environment by influencing more changes than originally expected.

The water column also faces multiple chances since the force it transfers through turbulence is being absorbed by the tidal farms. Waves would be directly affected by the reduced energy behind the water causing them to be weaker which too could destroy multiple ecosystems. There would also be an effect on the inter-tidal zones with there being less turbulence which many species like fish and crabs use for sustenance and survival. Other things like noise and electromagnetic fields also pose problems for the environment, but not to the scale that the last effects had on the geography and life in an ecosystem.

Effects on Ecology

The prime placement of tidal farms occurs in tidal barrages, marshes, lagoons and other intertidal bodies of waters most commonly home to migratory wetland birds. The placement of tidal farms raises the water levels to a point where the feeding areas of the birds are submerged. With the loss of feeding areas, mortality rates increase. Furthermore, studies were conducted by University of Exeter faculty for wetland birds in the Solway Firth, UK. These studies concluded that the effect of certain types of tidal farms are negligible. The impacts of tidal farms on water foul habitats in the Solway Firth were found to be relatively low, even for the largest case. The study highlighted that current installations with similar capacities as the simulations had a lower loss of intertidal area than that of its computer-generated counterpart. One conclusion reached was that future studies should focus on the vulnerability of an individual species, as compared to the effect on maritime birds as a whole.

The construction of tidal barrages within a bay, inlet, or estuary is one of the few cases where inter species relations were universally affected. The retention of tidal water alters the timetable of species which dominate the upper to intermediate shore habitats. Simultaneously the lower shore remains submerged for longer periods of time. The resulting forced ecosystem is disruptive to the majority of species present, with the exception of fish whom are tidal feeders. The extended periods of high tide allow for greater foraging opportunity, hence a growth in population can be observed in these cases.

The structures of larger tidal barrages alter the scour and deposition in their respective habitats. Scour and deposition refers to the movement and exchange of sediment along the floor of a water body. The U.S. Army Engineer Research and Development Center in Vicksburg, Mississippi stated that the interruption of natural sediment deposits directly led to an increased mortality rate of seabed grass, as the shoot could not properly grow in the altered seafloor. This would be even more devastating for benthic life which resides under the ocean floor since they are easily affected by changes in the flow over the bottom of the ocean. It would be hard to prevent this too since these organisms aren't easily relocated since they are so sensitive to changes in their atmosphere.

TIDAL BARRAGE

The Rance Tidal Power Station is a tidal barrage in France.

A tidal barrage is a dam-like structure used to capture the energy from masses of water moving in and out of a bay or river due to tidal forces.

Instead of damming water on one side like a conventional dam, a tidal barrage allows water to flow into a bay or river during high tide, and releases the water during low tide. This is done by measuring the tidal flow and controlling the sluice gates at key times of the tidal cycle. Turbines are placed at these sluices to capture the energy as the water flows in and out.

Tidal barrages are among the oldest methods of tidal power generation, with tide mills being developed as early as the sixth century. In the 1960s the 1.7 megawatt Kislaya Guba Tidal Power Station in Kislaya Guba, Russia was built.

Generating Methods

An artistic impression of a tidal barrage, including embankments, a ship lock, and caissons housing a sluice and two turbines.

The barrage method of extracting tidal energy involves building a barrage across a bay or river that is subject to tidal flow. Turbines installed in the barrage wall generate power as water flows in and out of the estuary basin, bay, or river. These systems are similar to a hydro dam that produces static head or pressure head (a height of water pressure). When the water level outside of the basin or lagoon changes relative to the water level inside, the turbines are able to produce power.

The basic elements of a barrage are caissons, embankments, sluices, turbines, and ship locks. Sluices, turbines, and ship locks are housed in caissons (very large concrete blocks). Embankments seal a basin where it is not sealed by caissons. The sluice gates applicable to tidal power are the flap gate, vertical rising gate, radial gate, and rising sector.

Only a few such plants exist. The first was the Rance Tidal Power Station, on the Rance river, in France, which has been operating since 1966, and generates 240MW. A larger 254MW plant began operation at Sihwa Lake, Korea, in 2011. Smaller plants include the Annapolis Royal Generating Station on the Bay of Fundy, and another across a tiny inlet in Kislaya Guba, Russia. A number of proposals have been considered for a barrage across the River Severn, from Brean Down in England to Lavernock Point near Cardiff in Wales.

Barrage systems are dependent on high civil infrastructure costs associated with what is in effect a dam being placed across estuarine systems. As people have become more aware of environmental issues, they have opposed barrages because of the adverse effects associated with changing a large ecosystem that is habitat for many varieties of species.

Energy calculations

The energy available from a barrage is dependent on the volume of water. The potential energy contained in a volume of water is:

$$E=\tfrac{1}{2}A\rho g h^2$$

where,

h is the vertical tidal range,

A is the horizontal area of the barrage basin,

ρ is the density of water = 1025 kg per cubic meter (seawater varies between 1021 and 1030 kg per cubic meter) and

g is the acceleration due to the Earth's gravity = 9.81 meters per second squared.

The factor half is due to the fact, that as the basin flows empty through the turbines, the hydraulic head over the dam reduces. The maximum head is only available at the moment of low water, assuming the high water level is still present in the basin.

Example Calculation of Tidal Power Generation

Assumptions:

- The tidal range of tide at a particular place is 32 feet = 10 m (approx).
- The surface of the tidal energy harnessing plant is 9 km^2 (3 km × 3 km)= 3000 m × 3000 m = 9×10^6 m^2.
- Density of sea water = 1025.18 kg/m^3.

Mass of the sea water = volume of sea water × density of sea water,

= (area × tidal range) of water × mass density

= $(9 \times 10^6\ m^2 \times 10\ m) \times 1025.18\ kg/m^3$

= 92×10^9 kg (approx)

Potential energy content of the water in the basin at high tide = ½ × area × density × gravitational acceleration × tidal range squared,

= $\frac{1}{2} \times 9 \times 10^6\ m^2 \times 1025\ kg/m^3 \times 9.81\ m/s^2 \times (10\ m)^2$

= 4.5×10^{12} J (approx)

Now we have 2 high tides and 2 low tides every day. At low tide the potential energy is zero. Therefore, the total energy potential per day = Energy for a single high tide × 2,

= $4.5 \times 10^{12}\ J \times 2$

= 9×10^{12} J

Therefore, the mean power generation potential = Energy generation potential / time in 1 day.

$= 9 \times 10^{12}$ J / 86400 s

= 104 MW

Assuming the power conversion efficiency to be 30%: The daily-average power generated = 104 MW * 30%

= 31 MW (approx)

Because the available power varies with the square of the tidal range, a barrage is best placed in a location with very high-amplitude tides. Suitable locations are found in Russia, USA, Canada, Australia, Korea, the UK. Amplitudes of up to 17 m (56 ft) occur for example in the Bay of Fundy, where tidal resonance amplifies the tidal range.

Tidal Barrage Generation

Tidal barrage generation using the tides is very similar to hydroelectric generation, except the water flows in two directions rather than just one. On incoming high tides, the water flows in one direction and fills up the tidal reservoir with sea water. On outgoing ebbing tides, the sea water flows in the opposite direct emptying it. As a tide is the vertical movement of water, the tidal barrage generator exploits this natural rise and fall of tidal waters caused by the gravitational pull of the sun and the moon.

The gravitational effects of the sun or the moon on the worlds oceans causes huge amounts of sea water to be directed towards the nearest coastline. The result of this movement of water is a rise in the sea level. In the open ocean, this rise is very small as there is a large surface area with deeper depths for it to flow into.

However, as the oceans water moves nearer towards the coastline, the sea level rises steeply especially around inlets and estuaries because of the upward sloping gradient of the sea bed. The effect of this sloping gradient is to funnelling the water into the estuaries, lagoons, river inlets and other such tidal "bottlenecks" along the coastline.

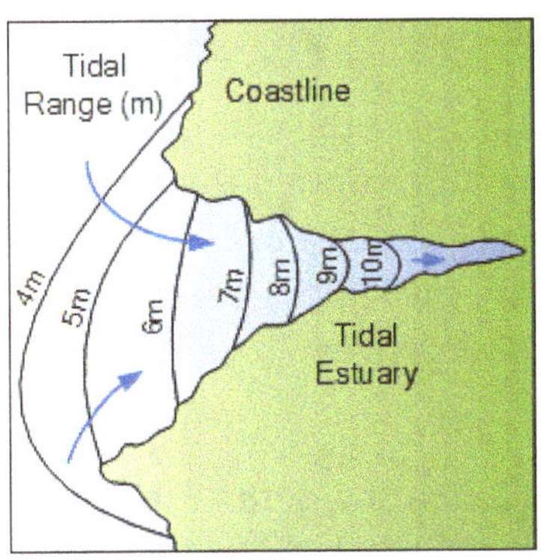

Effects of Tidal Flow in an Estuary.

The result of funnelling all of this water is that the height of the sea level once inside these inlets can increase vertically many metres every day as it is being pushed forward by the incoming sea water behind it as shown in the image. This increase in the sea level can create a tidal range of over ten metres in height in some estuaries and locations which can be exploited to generate electricity.

The tidal range is the vertical difference between the high tide sea level and the low tide sea level. The tidal energy extracted from these tides is potential energy as the tide moves in a vertical up-down direction between a low and a high tide and back to a low creating a height or head differential. A tidal barrage generation scheme exploits this head differential to generate electricity by creating a difference in the water levels either side of a dam and then passing this water difference through the turbines. The three main tidal energy barrage schemes that use this water differential to their advantage are:

- Flood Generation: The tidal power is generated as the water enters a tidal reservoir on the incoming Flood tide.
- Ebb Generation: The tidal power is generated as the water leaves a tidal reservoir on the Ebb flow tide.
- Two-way Generation: The tidal power is generated as the water flows in both directions in and out of the reservoir during both the Flood and the Ebb tides.

Tidal Barrage Flood Generation

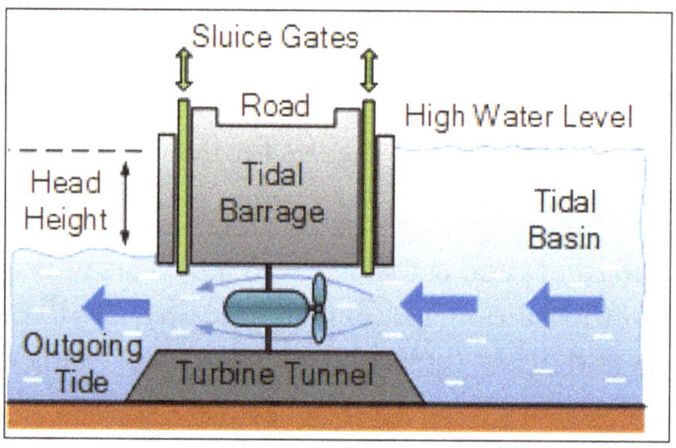

A Tidal Barrage Flood Generation uses the energy of an incoming rising tide as it moves towards the land. The tidal basin is emptied through sluice gates or lock gates located along the a section of the barrage and at low tide the basin is affectively empty. As the tide turns and starts to comes in, the sluice gates are closed and the barrage holds back the rising sea level, creating a difference in height between the levels of water on either side of the barrage dam.

The sluice gates to the entrances to the dams tunnels can either be closed as the sea water rises to allow for a sufficient head of water to develop between the sea level and the basin level before being opened generating more kinetic energy as the water rushes through, turning the turbines as it passes. Or may remain fully open, filling up the basin more slowly and maintaining the same water level inside the basin as out in the sea.

The tidal reservoir is therefore filled up through the turbine tunnels which spin the turbines generating tidal electricity on the flood tide and is then emptied through the opened sluice or lock gates on the ebb tide. Then a flood tidal barrage scheme is a one-way tidal generation scheme on the incoming tide with tidal generation restricted to about 6 hours per tidal cycle as the basin fills up.

The movement of the water through the tunnels as the tidal basin fills up can be a slow process, so low speed turbines are used to generate the electrical power. This slow filling cycle allows for fish or other sea life to enter the enclosed basin without danger from the otherwise fast rotating turbine blades. Once the tidal basin is full of water at high tide, all the sluice gates are opened allowing all the trapped water behind the dam to return back to the ocean or sea as it ebbs away.

Flood generator tidal power generates electricity on a incoming or flood tide, but this form of tidal energy generation is generally much less efficient than generating electricity as the tidal basin empties, called "Ebb Generation". This is because the amount of kinetic energy contained in the lower half of the basin in which flood generation operates is much less the kinetic energy present in the upper half of the basin in which ebb generation operates due to the effects of gravity and the secondary filling of the basin from inland rivers and streams connected to it via the land.

Tidal Barrage Ebb Generation

A Tidal Barrage Ebb Generation uses the energy of an outgoing or falling tide, referred to as the "ebb tide", as it returns back to the sea making it the opposite of the previous flood tidal barrage scheme. At low tide, all the sluice and lock gates along the barrage are fully opened allowing the tidal basin to fill up slowly at a rate determined by the incoming flood tide.

When the ocean or sea level feeding the basin reaches its highest point at high tide, all the sluices and lock gates are then closed entrapping the water inside the tidal basin (reservoir). This reservoir of water may continue to fill-up due to inland rivers and streams connected to it from the land.

As the level of the ocean outside the reservoir drops on the outgoing tide towards its low tide mark, a difference between the higher level of the entrapped water inside the tidal reservoir and the actual sea level outside now exists. This difference in vertical height between the high level mark and the low mark is known as the "head height".

At some time after the beginning of the ebb tide, the difference in the head height across the tidal barrage between the water inside the tidal reservoir and the falling tide level outside becomes sufficiently large enough to start the electrical generation process and the sluice gates connected to the turbine tunnels are opened allowing the water to flow.

When the closed sluice gates are opened, the trapped potential energy of the water inside flows back out to the sea under the enormous force of both gravity and the weight of the water in the reservoir basin behind it. This rapid exit of the water through the tunnels on the outgoing tide causes the turbines to spin at a fast speed generating electrical power.

The turbines continue to generate this renewable tidal electricity until the head height between the external sea level and the internal basin is too low to drive the turbines at which point the turbines are disconnected and the sluice gates closed again to prevent the tidal basin from over draining and effecting local wildlife. At some point the incoming flood tide level will again be at a sufficient level to open all the lock gates filling-up the basin and repeating the whole generation cycle over again as shown.

Power Generation during Ebb Tide

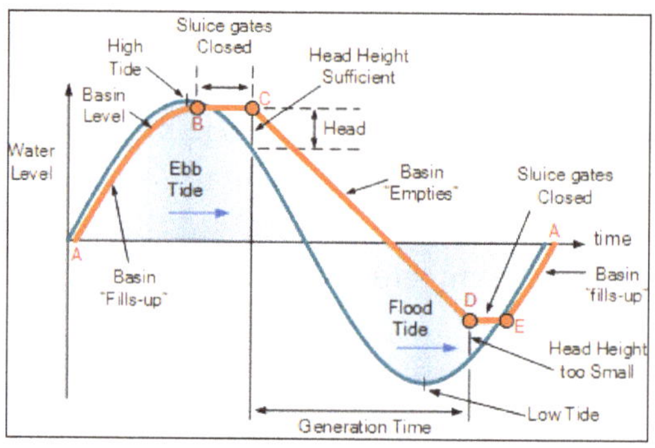

From the graph we can see that the tidal basin fills up between points E and B via A on the incoming flood tide. Once high tide is reached, the sluice gates are then closed between points B – C to stop the tidal basin from emptying. Therefore no power is generated between points E – C while the basin is filling up.

When there is sufficient head height between either side of the barrage, the sluice gates are opened at point C releasing the trapped water back to the sea and power generation starts. The tidal basin continues to empty driving the tidal generators until the level of water in the basin reaches point D.

At point D the head height of the sea water across the barrage is no longer adequate to drive the turbines so the sluice gates are closed and generation stops until the level of the sea water reaches point E again and the whole process starts over. Then Ebb Tidal Barrage Generation also known as outgoing generation, takes its name because the electrical generation only occurs as the tide ebbs or flows out.

At point E the sea level becomes sufficient to re-fill the basin and the sluice gates are opened. Therefore, the tidal reservoir is filled up through the sluice or lock gates on the flood tide and is emptied through the turbine tunnels which spin the turbines generating tidal electricity on the ebb tide. Then an ebb tidal barrage scheme is a one-way tidal generation scheme which operates between points C – D on the graph above.

Two-way Tidal Barrage Generation Scheme

We have seen above that both Flood Tidal Barrage and Ebb Tidal Barrage installations are "one-way" tidal generation schemes, but in order to increase the power generation time and therefore improve efficiency, we can use special double effect turbines that generate power in both directions. A Two-way Tidal Barrage Scheme uses the energy over parts of both the rising tide and the falling tide to generate electricity.

Two-way electrical generation requires a more accurate control of the sluice gates keeping them closed until the differential head height is sufficient in either direction before being opened. As the tide ebbs and flows, sea water flows in or out of the tidal reservoir through the same gate system. This flow of tidal water back and forth causes the turbine generators located within the tunnel to rotate in both directions producing electricity.

However, this two-way generation is in general less efficient than one-way flood or ebb generation as the required head height is much smaller which reduces the period over which normal one-way generation might have otherwise occurred. Also, bi-directional tidal turbine generators designed to operate in both directions are generally more expensive and less efficient than dedicated uni-directional tidal generators.

One way of improving the operating time and efficiency of a two-way tidal barrage scheme is to use individual one-way uni-directional tidal turbines inverted along the barrage. By controlling their individual sluice gates, one set of turbines can be made to operate on the flood tide between points E - B and the other set operate on the ebb tide between points C - D. While this method increases the total number of tidal turbines located along the tidal barrage, it has the advantage that the generation period is greatly extended.

Benefits

These types of dams can be used not only to produce renewable energy, but also to reduce flooding. As the global effects of climate change becomes more evident, the world is faced with the very real threat of rising sea levels. Tidal barrages can be used to prevent flooding of low lying areas - areas that already are at a high risk of flooding - by holding back or releasing water as necessary.

Furthermore, the Moon and Sun aren't going anywhere anytime soon (nor are their influence on the tidal forces) which means that we can easily produce this type of tidal power for as long as we need. This also means that tides are easily predictable and repetitive which allows us to know exactly when and how much power is attainable.

Drawbacks

Tidal barrages alter the flow of saltwater in and out of estuaries, which can alter the quality of the seawater and thus negatively impact and displace marine life in the area.

Furthermore, barrage systems have high development costs and it can be difficult to justify using so much capital especially when the environmental threats are taken into consideration.

The other, perhaps obvious, problem with tidal power is that you need tides. It is important to think of tidal power not as a one-off solution for the world's energy needs, but rather, to see it as a contributor to a low-carbon electricity generation portfolio.

Environmental Impact

The placement of a barrage into an estuary has a considerable effect on the water inside the basin and on the ecosystem. Many governments have been reluctant in recent times to grant approval for tidal barrages. Through research conducted on tidal plants, it has been found that tidal barrages constructed at the mouths of estuaries pose similar environmental threats as large dams. The construction of large tidal plants alters the flow of saltwater in and out of estuaries, which changes the hydrology and salinity and possibly negatively affects the marine mammals that use the estuaries as their habitat The La Rance plant, off the Brittany coast of northern France, was the first and largest tidal barrage plant in the world. It is also the only site where a full-scale evaluation of the ecological impact of a tidal power system, operating for 20 years, has been made.

French researchers found that the isolation of the estuary during the construction phases of the tidal barrage was detrimental to flora and fauna, however; after ten years, there has been a "variable degree of biological adjustment to the new environmental conditions."

Some species lost their habitat due to La Rance's construction, but other species colonized the abandoned space, which caused a shift in diversity. Also as a result of the construction, sandbanks disappeared, the beach of St. Servan was badly damaged and high-speed currents have developed near sluices, which are water channels controlled by gates.

Turbidity

Turbidity (the amount of matter in suspension in the water) decreases as a result of smaller volume of water being exchanged between the basin and the sea. This lets light from the Sun penetrate the water further, improving conditions for the phytoplankton. The changes propagate up the food chain, causing a general change in the ecosystem.

Tidal Fences and Turbines

Tidal fences and turbines, if constructed properly, pose less environmental threats than tidal barrages. Tidal fences and turbines, like tidal stream generators, rely entirely on the kinetic motion of the tidal currents and do not use dams or barrages to block channels or estuarine mouths. Unlike barrages, tidal fences do not interrupt fish migration or alter hydrology, thus these options offer energy generating capacity without dire environmental impacts. Tidal fences and turbines can have varying environmental impacts depending on whether or not fences and turbines are constructed with regard to the environment. The main environmental impact of turbines is their impact on fish. If the turbines are moving slowly enough, such as low velocities of 25-50 rpm, fish kill is minimalized and silt and other nutrients are able to flow through the structures For example, a 20 kW tidal turbine prototype built in the St. Lawrence Seaway in 1983 reported no fish kills Tidal fences block off channels, which makes it difficult for fish and wildlife to migrate through those channels. In order to reduce fish kill, fences could be engineered so that the spaces between the caisson wall and the rotor foil are large enough to allow fish to pass through Larger marine mammals such as seals or dolphins can be protected from the turbines by fences or a sonar sensor auto-braking system that automatically shuts the turbines down when marine mammals are detected.

Salinity

As a result of less water exchange with the sea, the average salinity inside the basin decreases, also affecting the ecosystem. "Tidal Lagoons" do not suffer from this problem.

Sediment Movements

Estuaries often have high volume of sediments moving through them, from the rivers to the sea. The introduction of a barrage into an estuary may result in sediment accumulation within the barrage, affecting the ecosystem and also the operation of the barrage.

Fish

Fish may move through sluices safely, but when these are closed, fish will seek out turbines and attempt to swim through them. Also, some fish will be unable to escape the water speed near a turbine and will be sucked through. Even with the most fish-friendly turbine design, fish mortality per pass is approximately 15% (from pressure drop, contact with blades, cavitation, etc.). Alternative passage technologies (fish ladders, fish lifts, fish escalators etc.) have so far failed to solve this problem for tidal barrages, either offering extremely expensive solutions, or ones which are used by a small fraction of fish only. Research in sonic guidance of fish is ongoing. The Open-Centre turbine reduces this problem allowing fish to pass through the open centre of the turbine.

Recently a run of the river type turbine has been developed in France. This is a very large slow rotating Kaplan-type turbine mounted on an angle. Testing for fish mortality has indicated fish mortality figures to be less than 5%. This concept also seems very suitable for adaption to marine current/tidal turbines.

Economics

Tidal barrage power schemes have a high capital cost and a very low running cost. As a result, a tidal power scheme may not produce returns for many years, and investors may be reluctant to participate in such projects.

Governments may be able to finance tidal barrage power, but many are unwilling to do so also due to the lag time before investment return and the high irreversible commitment. For example, the energy policy of the United Kingdom recognizes the role of tidal energy and expresses the need for local councils to understand the broader national goals of renewable energy in approving tidal projects. The UK government itself appreciates the technical viability and siting options available, but has failed to provide meaningful incentives to move these goals forward.

DYNAMIC TIDAL POWER

Dynamic tidal power is a technology that uses the difference between the potential energy and kinetic energy of tides. Long dams are built from coasts straight out into the sea or ocean—meaning that the tides in the areas where these systems might be implemented usually flow parallel to their respective coasts.During the movement of tides, water on one side of the dam is at a higher level than the other side. As this water flows through the dam it drives a series of turbines installed within the dam and generates electricity. Furthermore, these dams are designed with bi-directional turbines, which flip 180° after each tide in order to generate power both when the tide comes in and goes out. The added output from having bi-directional turbines is a huge advantage for these types of systems since they allow the power output to basically double.

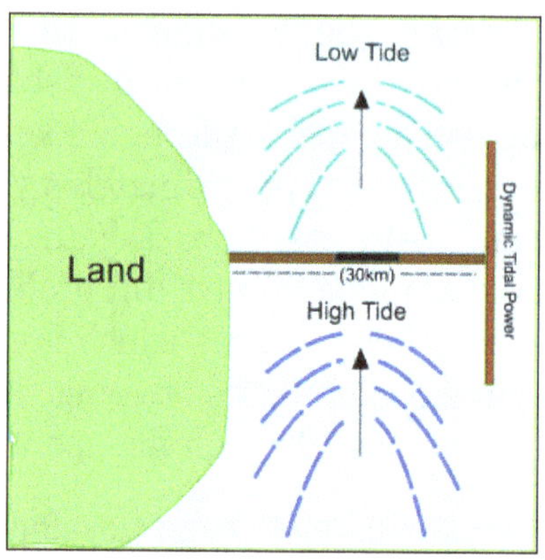

A representation of how Dynamic tidal power is used.

Dynamic tidal power generation remains in the developmental stages meaning that any data is based on models and known tidal behaviour. With that being said, these models tend to be highly advanced and have high confidence in these theoretical predictions.

A dynamic tidal power dam can be 30 to 60 km long and is typically built perpendicular to the coast, running relatively straight out into the ocean. This design does not enclose any area, thus allows more freedom for aquatic life without risk of being trapped. These dams generate a water level difference from one side of the dam to the other—known as 'head'. The head difference (difference in water level between both sides) is predicted to be able to reach up to a few meters. The higher the head difference will result in much higher power production.

Power Production

In 2015, Canada generated 670 terawatt-hours of electricity. This means that a single dynamic tidal power dam, with an 8 GW installed capacity (this is the theoretical maximum output of a power plant) and a capacity factor of 30% (this is the ratio of actual power output to theoretical power output) could generate almost 4% of Canada's total power output or 21 terawatt-hours annually.

Positives of Dynamic Tidal Power

Multiple Functions

These types of dams, by necessity, tend to be very long (30 to 60 km) which is significantly longer (and more expensive to build) than current coastal structures. However, a long dam can function with multiple purposes such as land reclamation, connections between islands, mainland coastal protection, and deep sea and liquefied natural gasports. The initial investment costs could easily be shared between these different industries, thus helping to lower the cost per kWh and create higher return investments.

Stable, Sustainable and Available

The moon (and subsequent tidal force) isn't going anywhere anytime soon which means that this type of tidal power will be around indefinitely. Furthermore, the tidal cycle is predictable and repetitive which allows us to know exactly when and how much power is obtainable.

Interestingly enough, since these dams are built perpendicular to the shore and the tide flows parallel, dynamic tidal power doesn't require huge tidal ranges. Therefore, these could be implemented in locations throughout the world, which is advantageous over other forms of tidal power (ex. tidal stream generators and tidal barrages) since they tend to require more intense tidal fluctuations in order to be economically viable.

Negatives of Dynamic Tidal Power

Lack of Proof

Unfortunately, due to the nature of the tides and the design of these dams, a full project must be built in order to guarantee predicted success. Building a miniature dam would be ineffective since these dams require a predicted minimum length of 30 km to ensure economic viability. Such a dam would require an enormous capital investment and if the design failed a lot of money would be lost. Furthermore, construction of a structure that extends 30 km or further into the ocean has never been attempted—the technology is available but the difficulty is extremely high.

Environmental Impact

The environmental impact of such a system could be significant. Unfortunately, in areas where any tidal power stations might be built happen to be some of the most densely populated ecosystems in the ocean. Changes in water levels might harm plant and animal life and alter the sea water composition. Also, turbines in dynamic tidal power stations move quickly, and if protection is not built in, marine animals can be caught in the blades. If aquatic life drops in the area, birds typically found there might migrate to different places. An entire major ecosystem could be significantly altered. Further research is needed in order to determine just how large of an impact such stations might have.

High Costs

It is difficult to determine the cost per MW for these types of power stations since none have been built yet. However, the costs are sure to be much higher than any other field of energy generation.

While the costs are high and the results as yet unknown, this technology may still have a place in the future electrical grid. As technology advances (which it will) and as climate change becomes more prevalent (which is virtually certain) there will be additional pressure to look for power with lower greenhouse gas emissions. Thus, the more that is known about alternative energy resources, the lower their costs. With these lower costs, transitions in power generation will be smoother.

Challenges

A major challenge is that the proof of DTP functioning can only be demonstrated by putting it in practice. Testing the concept of DTP at a small scale within a demonstration project, would not be effective, since almost no power would be yielded. Not even at a dam length of 1 km or so, because the DTP principle is such that the power generation capacity increases as the square of the dam length increases (both head and volume increase in a more or less linear manner for increased dam length, resulting in a quadratic increase in power generation). Economic viability is estimated to be reached for dam lengths of about 30 km.

Status of Technological Development

No DTP dam has ever been built, although all of the technologies required to build a DTP dam are available. Various mathematical and physical models have been conducted to model and predict the 'head' or water level differential over a dynamic tidal power dam. The interaction between tides and long dams has been observed and recorded in large engineering projects, such as the Delta Works and the Afsluitdijk in the Netherlands. The interaction of tidal currents with natural peninsulas is also well-known, and such data is used to calibrate numerical models of tides. Formulas for the calculation of added mass were applied to develop an analytical model of DTP. Observed water level differentials closely match current analytical and numerical models. Water level differential generated over a DTP dam can now be predicted with a useful degree of accuracy.

Some of the key elements required include:

- Bi-directional turbines (capable of generating power in both directions) for low head, high-volume environments. Operational units exist for seawater applications, reaching an efficiency of over 75%.
- Dam construction methods. This could be achieved by modular floating caissons (concrete building blocks). These caissons would be manufactured on shore and subsequently floated to the dam location.
- Suitable sites to demonstrate DTP. A pilot project of DTP could be integrated with a planned coastal development project, such as a sea bridge, island connection, deep sea port, land reclamation, offshore wind farm, etc., built in a suitable environment for DTP.

TIDAL LAGOON

A tidal lagoon is a power station that generates electricity from the natural rise and fall of the tides. Tidal lagoons work in a similar way to tidal barrages by capturing a large volume of water behind a man-made structure which is then released to drive turbines and generate electricity. Unlike a barrage, where the structure spans an entire river estuary in a straight line, a tidal lagoon encloses an area of coastline with a high tidal range behind a breakwater, with a footprint carefully designed for the local environment.

As the tide comes in (floods) the water is held back by the turbine wicket gates, which are used to control flow through the turbine and can be completely closed to stop water entering the lagoon. This creates a difference in water level height (head) between the inside of the lagoon and the sea. Once the difference between water levels is optimised, the wicket gates are opened and water rushes into the lagoon through the bulb turbines mounted inside concrete turbine housings in a section of the breakwater wall. As the water turns the turbines, electricity is generated.

The water in the lagoon then returns to closely match the same level as the sea outside. This process also happens in reverse as the tide flows out (ebbs) because the turbines are 'bi-directional' and so electricity can be generated from the incoming and outgoing tides. We can hold the tide within the lagoon for approximately 2.5 hours as the sea outside ebbs and the head builds.

As the tides rise and fall naturally, with no requirement for fuel, tidal power is truly renewable and, unlike other forms of renewable energy, is entirely predictable. As there are always two high and two low tides every day, tidal lagoons will generate electricity over four periods a day, every day of the year. As we hold the tides for 2.5 hours four times a day, we can generate power for up to 14 in every 24 hours. The height and time of the tides can be predicted years in advance to a high degree of accuracy, allowing the precise operation of the lagoon on each tidal cycle to be optimised well in advance.

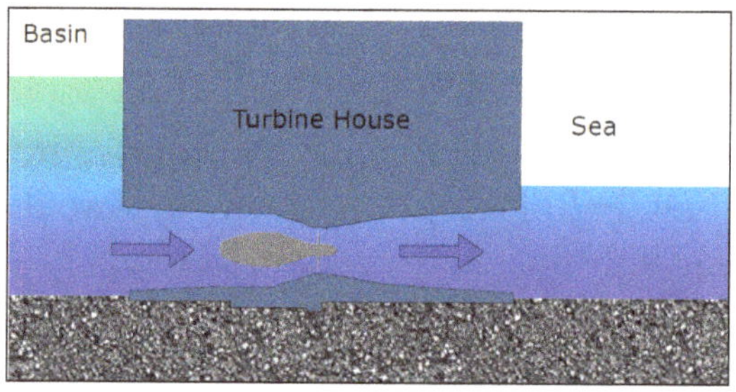

Generating on the flood tide.

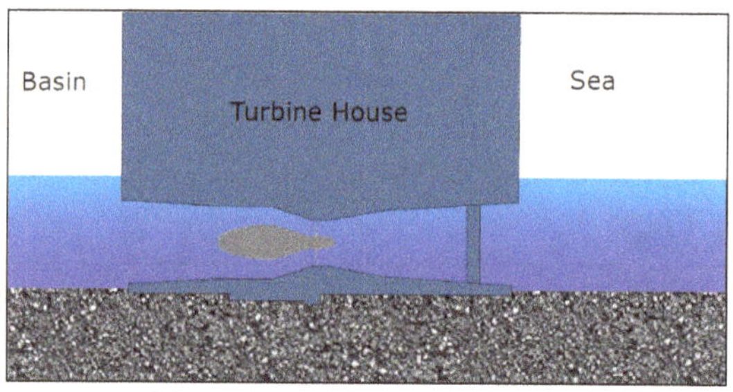

Holding period at high or low water.

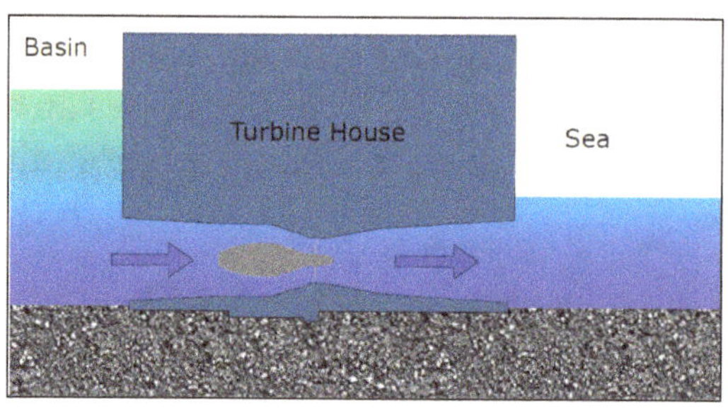

Generating on the ebb tide.

The Lagoon Wall

The primary purpose of the lagoon wall is to 'dam up' the ebb and of flood tide. To achieve this, the wall is required to have a very low permeability (leakage rate) to prevent the tide from escaping through the wall rather than flowing through the turbines. By constructing the central portion of the wall from a dense sand or gravel, and protecting it against the incoming waves and weather degradation with large armour rock, the wall will function as intended.

This type of wall is commonly known as a rubble mound breakwater with a sand core and is similar to many seen in coastal defence schemes and harbour breakwaters throughout the world. The wall will be constructed using proven, well established, marine construction methods. By partnering with leading global engineering consultancies, as well as consultation with specialist marine construction companies, we have ensured a robust, constructible and expert approach to our lagoon wall design.

The concrete structures housing the turbines and sluice gates are built so that the turbines are permanently submerged.

The size and shape of the lagoon depends on a lot of factors. As well as the tidal range, the water depth, sea conditions, environmental impacts and navigation all have a big influence on which locations are suitable to build a lagoon.

TIDE MILL

A tide mill is a water mill driven by tidal rise and fall. A dam with a sluice is created across a suitable tidal inlet, or a section of river estuary is made into a reservoir. As the tide comes in, it enters the mill pond through a one-way gate, and this gate closes automatically when the tide begins to fall. When the tide is low enough, the stored water can be released to turn a water wheel.

Tide mills are usually situated in river estuaries, away from the effects of waves but close enough to the sea to have a reasonable tidal range. Cultures have built such mills have existed since the Middle Ages, and some may date to the Roman period. A modern version of a tide mill is the electricity-generating tidal barrage.

Tidal mill at Olhão, Portugal.

Possibly the earliest tide mill in the Roman world was located in London on the River Fleet, dating to Roman times.

Three Mills, Stratford, one of the world's earliest recorded tide mills.

Since the late 20th century, a number of new archaeological finds have consecutively pushed back the date of the earliest tide mills, all of which were discovered on the Irish coast: A 6th-century vertical-wheeled tide mill was located at Killoteran near Waterford. A twin-flume, horizontal-wheeled tide mill, dating to c. 630, was excavated on Little Island in Cork. Alongside it, another tide mill was found that was powered by a vertical undershot wheel. The Nendrum Monastery mill from 787 was situated on an island in Strangford Lough in Northern Ireland. Its millstones are 830mm in diameter and the horizontal wheel is estimated to have developed 7/8HP at its peak. Remains of an earlier mill dated at 619 were also found at the site.

Three Mills, House Mill and Miller's House at low tide.

The earliest *recorded* tide mills in England are listed in the Domesday Book (1086). Eight mills are recorded on the River Lea (the site at Three Mills remains, with Grade I listed buildings and a small museum), as well as a mill in Dover harbour. By the 18th century, there were about 76 tide mills in London, including two on London Bridge.

Woodbridge Tide Mill, an excellent example, survives at Woodbridge, Suffolk, England. This mill, dating from 1170 and reconstructed in 1792, has been preserved and is open to the public. It was further restored in 2010 and re-opened in 2011 in full working order. It is the second working tide mill in the United Kingdom that is regularly producing flour. Carew Castle in Wales also has an intact tide mill, but it is not operating. The first tide mill to be restored to working order is Eling Tide Mill in Eling, Hampshire. Another example, now extant only in historic documents, is the mill in the hamlet of Tide Mills, East Sussex. Traces of a tide mill may be seen at Fife Ness, revealed through an archaeological survey.

Tidal mill at l'île de Bréhat.

A mediæval tide mill still operates at Rupelmonde near Antwerp, and there are several that have survived in the Netherlands.

At one time there were 750 tide mills operating along the shores of the Atlantic Ocean: approximately 300 in North America, including many in colonial Boston over a 150-year span. In addition, 200 have been documented in the British Isles, and 100 in France. The Rance estuary in France was also home to some of these mills.

By the mid-20th century, the use of water mills had declined dramatically. In 1938, an investigation by Rex Wailes discovered that of the 23 extant tidal mills in England, only 10 were still working by their own motive power. Of one at Beaulieu, H. J. Massingham wrote in the 1940s:

> "Part of the mill is built on piles into the river and is weatherboarded, while the rest of the building is a warm red brick roofed with lozenge-shaped and rounded tiles which I believe are called fish-tiles. All the interior is of wood - ladders, bins for the meal, floor-boarding, square pillars, beams, narrow passages, fittings, shaft rising to the first floor and all. So ramshackle is the arrangement of the props and supports that it is a wonder that the whole edifice does not tumble about the miller's ears like a pack of cards. The point is that it has stood in this way for something like six centuries, and that gives the explorer into its dusky depths a more penetrating notion of how the old builders could build, more than does a Gothic church or even a cathedral. The pulse and swing of the great wheel sets the whole building in an ague, but it will still be standing when all the flimsy excrescences of development between Beaulieu and Poole have fallen down."

Newer types of tidal power often propose construction of a dam across a large river estuary. Although hydroelectric power represents a source of renewable energy, each proposal tends to come under local opposition because of its likely adverse effect on coastal habitats. One proposal, which was developed in 1966, is the Rance barrage, which generates 250MW. Unlike historical tide mills, which could operate only on an ebb tide, the Rance barrage can generate electricity on both flows of the tide, or it can be used for pumped storage, depending on demand. A less intrusive design is a 1MW free-standing turbine, constructed in 2007 at Strangford Lough Narrows; this site is close to an historic tide mill.

References

- Tidal-power, entry: newworldencyclopedia.org, Retrieved 13 March, 2019
- Holzman, DC (2007). "Blue power: turning tides into electricity". Environ. Health Perspect. 115: A590–3. Doi:10.1289/ehp.115-a590. PMC 2137118. PMID 18087578
- Tidal-stream-generator, encyclopedia: energyeducation.ca, Retrieved 29 June, 2019
- "First power station to harness Moon opens - September 22, 2003". New Scientist. 2003-09-22. Retrieved 2013-04-28
- What-is-a-tidal-lagoon, tidal-technology: tidallagoonpower.com, Retrieved 30 January, 2019
- Charlier, Roger H. (December 2007). "Forty candles for the Rance River TPP tides provide renewable and sustainable power generation". Renewable and Sustainable Energy Reviews. 11 (9): 2032–2057. Doi:10.1016/j.rser.2006.03.015
- Tidal-barrage, tidal-energy: alternative-energy-tutorials.com, Retrieved 31 March, 2019
- Tidal-barrage, encyclopedia: energyeducation.ca, Retrieved 14 July, 2019

Permissions

All chapters in this book are published with permission under the Creative Commons Attribution Share Alike License or equivalent. Every chapter published in this book has been scrutinized by our experts. Their significance has been extensively debated. The topics covered herein carry significant information for a comprehensive understanding. They may even be implemented as practical applications or may be referred to as a beginning point for further studies.

We would like to thank the editorial team for lending their expertise to make the book truly unique. They have played a crucial role in the development of this book. Without their invaluable contributions this book wouldn't have been possible. They have made vital efforts to compile up to date information on the varied aspects of this subject to make this book a valuable addition to the collection of many professionals and students.

This book was conceptualized with the vision of imparting up-to-date and integrated information in this field. To ensure the same, a matchless editorial board was set up. Every individual on the board went through rigorous rounds of assessment to prove their worth. After which they invested a large part of their time researching and compiling the most relevant data for our readers.

The editorial board has been involved in producing this book since its inception. They have spent rigorous hours researching and exploring the diverse topics which have resulted in the successful publishing of this book. They have passed on their knowledge of decades through this book. To expedite this challenging task, the publisher supported the team at every step. A small team of assistant editors was also appointed to further simplify the editing procedure and attain best results for the readers.

Apart from the editorial board, the designing team has also invested a significant amount of their time in understanding the subject and creating the most relevant covers. They scrutinized every image to scout for the most suitable representation of the subject and create an appropriate cover for the book.

The publishing team has been an ardent support to the editorial, designing and production team. Their endless efforts to recruit the best for this project, has resulted in the accomplishment of this book. They are a veteran in the field of academics and their pool of knowledge is as vast as their experience in printing. Their expertise and guidance has proved useful at every step. Their uncompromising quality standards have made this book an exceptional effort. Their encouragement from time to time has been an inspiration for everyone.

The publisher and the editorial board hope that this book will prove to be a valuable piece of knowledge for students, practitioners and scholars across the globe.

Index

www.ingramcontent.com/pod-product-compliance
Lightning Source LLC
LaVergne TN
LVHW080611210926
848485LV00003B/137